Viktor M. Kureichik, Sergey P. Malioukov, Vladimir V. Kureichik,
and Alexander S. Malioukov

Genetic Algorithms for Applied CAD Problems

Studies in Computational Intelligence, Volume 212

Further volumes of this series can be found on our homepage: springer.com

Vol. 192. Agus Budiyono, Bambang Riyanto and Endra Joelianto (Eds.)
Intelligent Unmanned Systems: Theory and Applications, 2009
ISBN 978-3-642-00263-2

Vol. 193. Raymond Chiong (Ed.)
Nature-Inspired Algorithms for Optimisation, 2009
ISBN 978-3-642-00266-3

Vol. 194. Ian Dempsey, Michael O'Neill and Anthony Brabazon (Eds.)
Foundations in Grammatical Evolution for Dynamic Environments, 2009
ISBN 978-3-642-00313-4

Vol. 195. Vivek Bannore and Leszek Swierkowski
Iterative-Interpolation Super-Resolution Image Reconstruction: A Computationally Efficient Technique, 2009
ISBN 978-3-642-00384-4

Vol. 196. Valentina Emilia Balas, János Fodor and Annamária R. Várkonyi-Kóczy (Eds.)
Soft Computing Based Modeling in Intelligent Systems, 2009
ISBN 978-3-642-00447-6

Vol. 197. Mauro Birattari
Tuning Metaheuristics, 2009
ISBN 978-3-642-00482-7

Vol. 198. Efrén Mezura-Montes (Ed.)
Constraint-Handling in Evolutionary Optimization, 2009
ISBN 978-3-642-00618-0

Vol. 199. Kazumi Nakamatsu, Gloria Phillips-Wren, Lakhmi C. Jain, and Robert J. Howlett (Eds.)
New Advances in Intelligent Decision Technologies, 2009
ISBN 978-3-642-00908-2

Vol. 200. Dimitri Plemenos and Georgios Miaoulis *Visual Complexity and Intelligent Computer Graphics Techniques Enhancements,* 2009
ISBN 978-3-642-01258-7

Vol. 201. Aboul-Ella Hassanien, Ajith Abraham, Athanasios V. Vasilakos, and Witold Pedrycz (Eds.)
Foundations of Computational Intelligence Volume 1, 2009
ISBN 978-3-642-01081-1

Vol. 202. Aboul-Ella Hassanien, Ajith Abraham, and Francisco Herrera (Eds.)
Foundations of Computational Intelligence Volume 2, 2009
ISBN 978-3-642-01532-8

Vol. 203. Ajith Abraham, Aboul-Ella Hassanien, Patrick Siarry, and Andries Engelbrecht (Eds.)
Foundations of Computational Intelligence Volume 3, 2009
ISBN 978-3-642-01084-2

Vol. 204. Ajith Abraham, Aboul-Ella Hassanien, and André Ponce de Leon F. de Carvalho (Eds.)
Foundations of Computational Intelligence Volume 4, 2009
ISBN 978-3-642-01087-3

Vol. 205. Ajith Abraham, Aboul-Ella Hassanien, and Václav Snášel (Eds.)
Foundations of Computational Intelligence Volume 5, 2009
ISBN 978-3-642-01535-9

Vol. 206. Ajith Abraham, Aboul-Ella Hassanien, André Ponce de Leon F. de Carvalho, and Václav Snášel (Eds.)
Foundations of Computational Intelligence Volume 6, 2009
ISBN 978-3-642-01090-3

Vol. 207. Santo Fortunato, Giuseppe Mangioni, Ronaldo Menezes, and Vincenzo Nicosia (Eds.)
Complex Networks, 2009
ISBN 978-3-642-01205-1

Vol. 208. Roger Lee, Gongzu Hu, and Huaikou Miao (Eds.)
Computer and Information Science 2009, 2009
ISBN 978-3-642-01208-2

Vol. 209. Roger Lee and Naohiro Ishii (Eds.)
Software Engineering, Artificial Intelligence, Networking and Parallel/Distributed Computing, 2009
ISBN 978-3-642-01202-0

Vol. 210. Andrew Lewis, Sanaz Mostaghim, and Marcus Randall (Eds.)
Biologically-Inspired Optimisation Methods, 2009
ISBN 978-3-642-01261-7

Vol. 211. Godfrey C. Onwubolu (Ed.)
Hybrid Self-Organizing Modeling Systems, 2009
ISBN 978-3-642-01529-8

Vol. 212. Viktor M. Kureichik, Sergey P. Malioukov, Vladimir V. Kureichik, and Alexander S. Malioukov
Genetic Algorithms for Applied CAD Problems, 2009
ISBN 978-3-540-85280-3

Viktor M. Kureichik, Sergey P. Malioukov,
Vladimir V. Kureichik, and Alexander S. Malioukov

Genetic Algorithms for Applied CAD Problems

Springer

Viktor M. Kureichik
The Taganrog Technological
Institute of the South Federal University
Department of Electronic Apparatuses Design
Nekrasovsky st., 44
Taganrog, GSP-17A
Russia, 347928
E-mail: kur@tsure.ru

Vladimir V. Kureichik
The Taganrog Technological
Institute of the South Federal University
Department of Electronic Apparatuses Design
Nekrasovsky st., 44
Taganrog, GSP-17A
Russia, 347928
E-mail: vkur@tsure.ru

Sergey P. Malioukov
The Taganrog Technological
Institute of the South Federal University
Department of Electronic Apparatuses Design
Nekrasovsky st., 44
Taganrog, GSP-17A
Russia, 347928
E-mail: malyukov@fep.tsure.ru

Alexander S. Malioukov
The Taganrog Technological
Institute of the South Federal University
Department of Electronic Apparatuses Design
Nekrasovsky st., 44
Taganrog, GSP-17A
Russia, 347928
E-mail: kes@fep.tsure.ru

ISBN 978-3-642-24257-1 e-ISBN 978-3-540-85281-0

DOI 10.1007/978-3-540-85281-0

Studies in Computational Intelligence ISSN 1860949X

Softcover reprint of the hardcover 1st edition 2009

Typeset & Cover Design: Scientific Publishing Services Pvt. Ltd., Chennai, India.

Printed in acid-free paper

9 8 7 6 5 4 3 2 1

springer.com

Foreword

The development of intelligent systems connecting the human brain and computer technologies represents one of the most important problems of the 21st century. Therefore, analytical methods of data mining of computer databases are being developed.

Intellectual behavior of technical objects as well as the biological ones is defined by their structure, architecture and a general organization first of all.
A purposeful direction can be defined as an intelligent behavior. It consists in finding the best ways to obtain some purpose by means of trial-and-error and study methods. These purposes are different for each category but all of them assume the control object adaptation to unpredictable changes of their characteristics in time.

The intelligent behavior of technical objects can be defined as simulation of some important functions of biological systems.
New perspective technologies of genetic search and evolution simulation represent the kernel of this book. The authors wanted to show how these technologies are used for practical problems solution.

This monograph is meant for specialists of computer aided design (CAD), intelligent information technologies in science, biology, economics, sociology and others. It may be used by post-graduate students and students of specialties connected to systems theory and system analysis, information science, optimization, operations rcscarch, ctc.

Professor A.P. Eremeev
Laureate of the President Prize in the field of Education
Chairman
Applied Mathematics and Information Science Department
Power Engineering Institute
Technical University
Moscow, Russia

Preface

The development of intellectual systems connecting the human brain and computer technologies represents one of the most important problems of the 21st century. Therefore analytical methods of data mining of computer databases are being developed.

Intellectual behavior of technical objects as well as the biological ones is defined by their structure, architecture and general organization first of all.

Purposeful direction can be defined as intellectual behavior. It consists in finding the best ways to obtain some purpose by means of trial-and-error and study methods. These purposes are different for each category but all of them suppose the control object adaptation to unpredictable changes of their characteristics in time.

The intellectual behavior of technical objects can be defined as simulation of some important functions of biological systems.

New perspective technologies of genetic search and evolution simulation represent the kernel of this book. The authors wanted to show how these technologies are used for practical problems solution.

This monograph is devoted to specialists of CAD, intellectual information technologies in science, biology, economics, sociology and others. It may be used by post-graduate students and students of specialties connected to the systems theory and system analysis methods, information science, optimization methods, operations investigation and solution-making.

Chairman
of the Applied Mathematics
and Information Science Department
of Power Engineering Institute
(Technical University), Moscow, Russia,
Laureate of the President Prize in the field of Education,
Doctor of Science, Professor
A.P. Eremeev

Kureichik V.M., Doctor of Engineering Science, Professor, Member of the Russian Academy of Natural and Engineering Sciences (RANS) and Senior Member of IEEE (USA), outstanding scientist in the area of development of mathematical models for design of the large and very large integrated circuits, firmware of intelligent computer aided design systems, founder and leader of the scientific school of graph theory, recognized in Russia and abroad, in genetic algorithms, intellectual evolution simulation, intelligent CAD systems development.

Phone: +7 863 4 311487,
+7 863 4 393260
E-mail: kur@tsure.ru

Malioukov S.P., Doctor of Engineering Science, Professor, Corresponding Member of the RANS, well-known expert in development of film magnetic heads, video heads with high-performance data carrier, multitrack magnetic heads, magneto-resistive heads, pressure sensors, inorganic vitreous dielectrics. Published more than 100 articles on physics and microelectronics. Author of 14 certified inventions.

Phone: +7 863 4 371603,
E-mail: malyukov@fep.tsure.ru

Kureichik V.V., Doctor of Engineering Science, Professor, Head of the CAD Department of the Technological Institute of the South Federal University, Taganrog, Russia; Corresponding Member of the Russian Academy of Engineering Science.

Phone: +7 863 4 383451,
+7 863 4 371651
E-mail: vkur@tsure.ru

Malioukov A.S., M. Sc. Department of Computer Science, Abo Academy University Finland. Specialist in development and use of optimization algorithms, adoption of existing theory to various technical applications. Developer of new intelligent CAD and heuristic software systems.

E-mail: cam@ezhe.ru

Contents

1 General Questions of Automated Design and Engineering............ 1
1.1 The Stages of Automated Design and Engineering.................... 1
1.2 The Problems of Computer Aided Design............................ 4
1.3 Review of the Modern CAD Systems.................................... 12
1.3.1 Classification of the CAD Systems.............................. 14
1.3.2 Universal CAD Systems.. 15
1.3.3 Main Functional Capabilities of Advanced CAD Systems... 16
1.4 Problem Statement of the DSD Magnetic Head Design Automation.. 19
1.5 Conclusions... 22

2 Evolutionary Models of Decision Making................................ 23
2.1 Evolutionary Methods and Genetic Algorithms in Artificial Systems.. 23
2.2 The Non-standard Genetic Search Architectures...................... 36
2.2.1 Development of the Evolutionary Adaptation Block.......... 36
2.2.2 Future-Oriented Technologies of Genetic Search for Solving the Optimization Problems............................ 51
2.3 Optimization Problems of Decision-Making on Graphs.............. 61
2.3.1 The Analysis of Genetic Algorithms of Graph Partitioning on the Basis of Artificial System Models...................... 61
2.3.2 The Genetic Approach to the Placement of the Graph Nodes.. 74
2.3.3 Solving the Traveling Salesman Problem by Simulated Evolution.. 84
2.3.4 Greedy Heuristics for Solving the Problems of Coloring, Isomorphism, Construction of Cliques and Independent Sets of Graphs.. 91
2.4 Construction and Analysis of the Evolutionary Algorithm for Finding Graph Pair Combinations.. 107
2.4.1 Combined Genetic Algorithm for Finding Pair Combinations.. 110
2.5 Conclusion.. 114

3 Algorithms and Models of ECE Module Connections Routing......... 115
3.1 The Statement of the Routing Problem and a Classification of Algorithms........ 115
3.2 Wire Connections Routing........ 120
3.3 Mathematical Model of Scheme Planarity and Its Flat Layout....... 125
3.4 Distributing Connections in Layers........ 131
3.5 Routing Algorithms 134
3.6 Improvement of Composition, Allocation and Routing Results Based on the Method of Equivalent Logical Function Translation... 143
3.7 Algorithms and Models for Design of Topology of Integrated Circuits and Microprocessors........ 148
3.8 Topological and Metric-Topological Approaches to LSI Design.... 157
3.9 Synthesis of the Metric-Topological Model for Array LSI Topological Outline........ 163
3.10 Striping of Combined Topology 167
3.11 Description Language and Control of LSI Scheme Topology....... 176
3.12 Questions of Hardware Support of VLSI CAD Software............ 182
3.13 Design of LSI Topology Using Silicon Compilers........ 186
3.14 Conclusion........ 191

4 Development of Genetic Algorithms for Finding Optimal Magnetic Head Parameters and Characteristics 193
4.1 Genetic Algorithm with Multiple Genotype........ 194
4.1.1 Finding Optimal Parameters of Magnetic Heads........ 194
4.1.2 Presentation of Genetic Material........ 195
4.1.3 The Solution Encoding Methods........ 196
4.1.4 The Target Function........ 197
4.1.5 Genetic Operators and Algorithm Structure........ 197
4.1.6 Theoretical Evaluation of the Algorithm........ 199
4.2 Dynamic Genetic Algorithm........ 201
4.2.1 The Purpose of Development of Dynamic Algorithm......... 201
4.2.2 The Structure of Presentation of Genetic Material........ 202
4.2.3 The Method of Encoding Solution........ 202
4.2.4 Target Function........ 203
4.2.5 Genetic Operators and Algorithm Structure........ 203
4.2.6 Theoretical Evaluation of the Algorithm........ 209
4.3 Conclusions........ 209

5 Experimental Investigation of Algorithms Developed........ 211
5.1 The Purpose of Experimental Investigation........ 211
5.2 Investigation of Genetic Algorithm with Multiple Genotype......... 212
5.2.1 Definition of Optimal Parameters........ 212
5.2.2 Space and Time Complexity........ 214
5.2.3 Comparative Characterization........ 215
5.3 Investigation of Dynamic Genetic Algorithm........ 215
5.3.1 Selection of Algorithm Parameters........ 215

5.3.2 Space and Time Complexity.. 219
5.3.3 Comparative Characterization.................................... 220
5.4 Conclusions.. 223

Final Conclusions.. 225
References.. 227

Abbreviations

ADP	automated design procedure
AI	artificial intelligence
AIS	artificial intelligence system
ART	algorithm running time
AS	artificial system
BB	building block
BEA	block of evolutionary adaptation
CAD	computer aided design
CAE	computer aided engineering
CAM	computer aided manufacturing
CD	connections diagram
CF	commutation field
CNC	computer numeric control
CO	crossover operator
CS	conflict situation
DB	database, databank
DCT	directed covering tree
DD	discrete device
DE	evolution on Darwin
DGA	dynamic genetic algorithm
DL	design language
DM	a method of dichotomy
DMP	decision-making person
DMSS	decision-making support system
DP	dialogue processor
DSD	data storage device
DTWF	discrete topological working field
DWF	discrete working field
ECE	electronic computing equipment
EE	electronic equipment

ES	expert system
FM	Fibonacci method
FN	Fibonacci numbers
FS	fractal set
GA	genetic algorithm
GAFA	genetic algorithm of formula approximation
GAMG	genetic algorithm with multiple genotype
GC	graph coloring
GCO	greedy crossing-over operator
GH	greedy heuristic
GIR	graph isomorphism recognition
GO	genetic operator
GS	genetic search
GSMS	genetic search management system
LSI	large-scale integrated circuit
HM	homeostatic management
IC	integrated circuit
IO	inversion operator
IPP	integer programming problem
IS	independent subset
KB	knowledge base
LE	evolution on Lamarck
LS	linguistic software
LTEC	linear thermal expansion coefficient
MD	micro-discrete
MH	magnetic head
MIS	metal-isolator-semiconductor
MM	mathematical model
MO	mutation operator
MRD	magnetic recording device
MRE	magneto-resistive element
MS	mathematical software
MTS	methodical software
NS	natural system
OT	optimization task
PTMRH	planar single-turn thin-film magneto-resistive head
RF	recombination function

RO	removal operator
RPO	reproduction operator
RS	random search
SA	simulated annealing
SC	silicon compiler
SCT	shortest covering tree
SET	synthetic theory of evolution
SGA	simple genetic algorithm
SO	segregation operator
SOM	statistical optimization method
SP	Steiner point
ST	Steiner tree
SW	Software
TF	target function
TO	translocation operator
TRE	natural system
TSP	traveling salesman problem
VLSI	very large-scale integrated circuit

Chapter 1
General Questions of Automated Design and Engineering

1.1 The Stages of Automated Design and Engineering

In a general case, the process of electronic equipment (EE) automated design, like design of all other discrete devices (DD), consists of three main stages: systems engineering, circuit engineering and design.

The first stage includes the system and structure design. The second stage - schematic design - consists of simulation, logical design, control and diagnostic test building. The third stage includes the engineering and technological design.

A diagram presenting one of possible schemes of the EE automated design process is shown in Fig. 1.1.

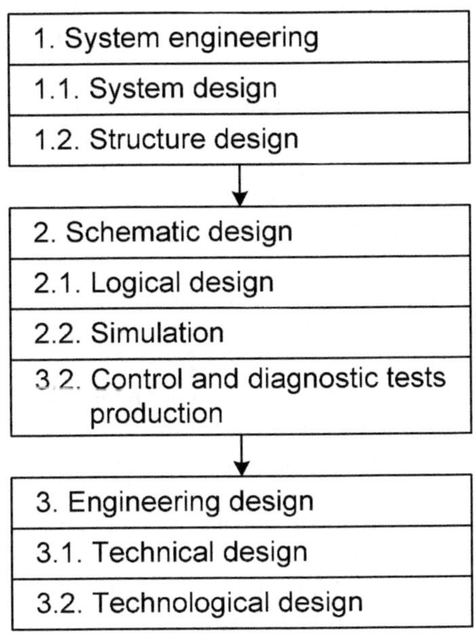

Fig. 1.1. EE automated design structural diagram

Let us describe the stages. During system design, the concepts and methods of system analysis are used. On the basis of multiple factors the fundamental analysis of EE development requirements specification is performed and then the decision is taken concerning the building methods and the ways of carrying out the calculation process.

During the structure design stage, the general structural diagram of the EE and the algorithms for separate operations are developed. For selecting the structure it is necessary to consider manufacturability and reliability requirements, and the possibility of using the uniform and quasi-uniform standard components.

The systems engineering stage generally remains a non-formalized process. The creative abilities of engineers are the main tool at this stage. Computer reviews the solution variants, accepted by the designer, and selects the optimal ones. At this stage, one uses special languages and formal methods of generating the calculation process variants for the source task.

V.M. Kureychik et al.: Genetic Algorithms for Applied CAD Problems, SCI 212, pp. 1–22.
springerlink.com

During the schematic design stage the logical and computational abilities of computers are widely used. The purpose of the logical EE design is the automated, formalized, abstractive and structural synthesis of components, selected during structure design, with appropriate correspondence between the initial task and the final result being checked. In terms of theory there are significant achievements in this domain: the control and special-type operational devices are synthesized automatically. In practice, the automation of design of the logical schemes requires having a solution to a large number of problems. Some of them are: development of effective languages for initial tasks description, of the structure design languages, of algorithms for construction of formal models of devices, etc. During logical design, the main optimization criteria are: minimization of logical unit type numbers, maximization of the logical unit reproducibility, ability of efficient simulation and diagnostics of the schemes, maximum consideration of engineering and technological design requirements. The problem of simulation consists in building of connection maps for logical signals, checking of time correlations during the passage of input signals, analysis of correspondence of functional schemes to definite Boolean functions. Simulation can, in general, be either physical or mathematical. For the EE schemes, mathematical simulation is more important, as usage of complex integrated circuits usually makes physical simulation impossible.

Note that according to the Soviet all-Union State Standard 17021—75, currently valid in Russia, we will assume that an *integrated circuit* is a microelectronic device that performs a definite function of signal conversion and processing and has high packaging density of electrically connected elements and chips; from the point of view of test requirements and acceptance it is regarded as a single whole.

An integrated circuit of the 1st integrity level (IC1) is a chip that contains up to 10 elements and components altogether.

An integrated circuit of the 2nd integrity level (IC2) is a chip that contains from 10 to 100 elements and components altogether.

An integrated circuit of the 3rd integrity level (IC3) is a chip that contains from 100 to 1 000 elements and components in total.

An integrated circuit of the 4th integrity level (IC4) is a chip that contains from 1 000 to 10 000 elements and components in total.

An integrated circuit of the 5th integrity level (IC5) is a chip that contains from 10 000 to 100 000 elements and components in total.

An integrated circuit of the 6th integrity level (IC6) is a chip that contains altogether between 100 000 and 1 000 000 elements and components.

A *highly-integrated chip* is an integrated circuit that contains 500 and more elements, produced by bipolar technology, or 1000 and more elements produced via MIS technology.

Microprocessor is a software programmable device that performs digital data processing and its control; it is based on one or several highly-integrated chips.

An i*ntegrated circuit element* is a part of IC that is not designated as a separate product and acts like some radio component (transistor, diode, etc.).

An *integrated circuit component* is a part of IC that acts as some radio element and can be designated as a separate product.

Control and diagnostics are developed at the simulation sub-stage. The methods for constructing the hardware control schemes are defined, test service systems are developed, and the required necessary conditions are defined in order to select the minimal reparable unit. This is connected with maximization of reliability of the elements used and of the possibility of replacement by typical elements in devices.

The functional schemes obtained after logical synthesis and simulation are the input information for the engineering (technical, assembling, physical) design. In this phase the following problems must be solved: covering of the functional schemes with the cells from a definite set, i.e. conversion to the electrical schematic diagram of the device; scheme element composition in terms of typical replacement elements (TRE)—reparable constructive units, panels, etc.; allocation of elements in the units according to different criteria; allocation of circuits in layers—multilayer or two-layer routing and control of correctness of the obtained topology.

The purpose of technological design is to produce in an automated manner the technological documents, to develop control algorithms for the peripheral devices producing design documents, and to develop methods of automatic production of photomaps, the guiding documents in the production system.

One of the most important problems consists, therefore in engineering design automation.

In this book the questions of methods, algorithms and automated development of the design systems are examined using modern mathematic methods. Design is based on mathematical description of the design problems in a formal language, formulation of theorems and algorithms, system structure, writing programs in the algorithmic language, and obtaining respective solutions with universal or specialized computers, their output being directed to workstations and other equipment.

For most of the design problems the formal fragmentation of the search process often gets difficult. The design problems, formulated in set-theoretical approach, usually lead to problems that can be solved only by enumeration of large number of variants. That is why it is important to find efficient methods of simplifying enumeration. An important role is played in this task by the issue of formal description of the informally stated problems, methods of their division into separate steps, organization of procedures of search for the design variants that would be optimal in various aspects.

The use of the apparatus of graph theory becomes increasingly widespread nowadays in solving problems of automated design and engineering. This is due to the fact that the language of graph theory is adequate in many cases to the subject of design, describing its aspects in a natural way, and at the same time allowing for an abstraction from certain material features and for operating with abstract models. This makes it possible to build mathematically-based design algorithms, to find simple and high-quality solutions, and to use computer rationally and efficiently. Finding of exact solutions to high-dimensional design problems involves enumeration of a large number of variants, and is difficult even for computers. Therefore, this work, along with the precise design methods, based on operations research methodology, examines the direction search algorithms that do not provide optimal solutions, but allow for obtaining results suitable for practical use.

1.2 The Problems of Computer Aided Design

Main problems in creation of computer aided design (CAD) systems are: improvement of design quality and creation of tools to provide solutions to the fundamentally new problems, resulting from technological progress.

In a general case, a *computer aided design system*, or an engineering design system is a set of algorithms with a controller, implemented as set of programs, assembled in packages, libraries or units; and automated workstations that include equipment, required for the engineering documentation production. An *ideal CAD system* assumes such a sequence of operations that the requirements specification, formulated by the designer, is fully processed using computer. The software system defines the sequencing and hence the ordering of implementation of the separate stages. At the output, computer produces a model of the topology of device, as the documentation for the automated control system of the technological process.

However, even the most modern computers cannot replace the designer; they only are able to help him by carrying out non-creative, routine operations. Therefore, at present, interactive "human-machine" systems are most widespread, working in the mode of dialogue with the designer. They are most efficient during analysis and solving of the combinatorial and logical problems for the stage of scheme design. Interactive systems should be organized so as to allow for optimal combination of the automated design process with the instructions from the designer, who creatively guides the design process. Next, a not less important factor is constituted by the definition of the application area and selection of design methodology. The importance of this factor is connected with inefficiency of universal algorithms when used under different circumstances. Hence, it is advisable to include in the system the controller programs that will manage other programs. The presence of the controller will also allow for solving of the following important issues in system organization: the possibility of a free "entry" into the system at all the design stages in order to control the intermediate results; the possibility of using the software as packages and/or as separate programs; and organization of the most rational sequence of the design stages.

Fulfillment of the requirements listed above becomes necessary during the step-by-step realization of the design process. In this context, system efficiency will depend largely upon the reliability and mutual fit of the separate design stages. This is achieved through unification of input and output information, common data storage methods, computer memory assignment, etc. A significant place in CAD organization is assigned the selection of the algorithmic language that should be able to describe the initial input information and remain simple enough for the designer. Note that the decisive factor for CAD is quantitative or qualitative benefit from automation that significantly exceeds the required advanced labor input. A developed system should have high viability, i.e. ability of accommodating the changes in optimization criteria, of extending the program library, of linking with other design tools and automated design processes.

At present time, CAD systems are being developed in two directions: on the one hand, mini- and microcomputers, as well as microprocessors, with direct designer collaboration are widely used. On the other hand, automated design systems, based on the multiprocessor computational structures without human participation are built. It is mostly understood that the second direction is the one with bigger perspectives. In both directions, though, there remain the questions of algorithmic optimization, formalization of the design problems, presentation of information in the computer, organization of program libraries, etc.

An automated design system should be able to: automatically store the data about the device designed; ensure the sequential expansion and improvement of the system, assure an active connection between the designer and the system; operate the optimally interchangeable design algorithms, provide for system specialization in the EE design for any level of the IC integrity; increase the system power through the usage of multiprocessor structures and peripheral devices; dock with special automates (various output devices, plotters, etc.); produce technological and design documentation.

The essential requirement for information allocation in computer memory is free access to data, i.e. the developer should be able to quickly examine all available parameters and select the required ones at all the design stages.

Not less important is correct development of the *design language* (DL), i.e. language for presentation and conversion of object descriptions during design. According to the Soviet all-Union State Standard 22487—77, valid currently in Russia, there are: the input design language for project statement; the basic design language for representation of additional information to the initial design object description, involving description of design solutions, design procedures and their sequences; and the output design language for representation of any design solution, including design results in the form that complies with requirements as to the use of the solution.

Correctness of selection of algorithms is one of factors defining economic efficiency of the CAD system use. This means that work is needed directed at further improvement of mathematical, informational, technical, linguistic, methodical, organizational, and software aspects of CAD.

Mathematical software (MS) of automated design (MSAD) is the entirety of mathematical methods, models and design algorithms required for solving the design problem.

*Software supplement (*SS) of automated design (SSAD) is the entirety of compiled programs represented in a definite form. Correspondingly, application package is the whole of the compiled programs, presented in a definite form, required for realization of the design procedure. The part of SSAD, meant for design control, is called operating system (OS) of automated design.

Informational software of automated design is the entirety of data, represented in definite form, required for CAD realization.

A part of the CAD information supplement includes automated databanks, consisting of the database (DB) and the database control systems. Automated databanks are created as service subsystems of CAD and are meant for automated supply of CAD subsystem with required data.

The automated databank is controlled by the experts. It provides for the usage integrity, correctness, efficiency, and functional capabilities.

Linguistic software (LS) of automated design is entirety of languages, represented in a definite from, including terms and definitions, natural language formalization rules, text compression and expansion methods, required for automated design.

Methodological supplement (MTS) of automated design is the whole set of documents that define composition, selection and exploitation rules for the design facilities.

Organizational supplement (ORS) of automated design is the whole set of documents that define composition of design organization and its subdivisions, connections between the subdivisions, their functions, as well as the form of presentation of the design results, and the order of examination of the design documents, required for automated design.

Then, the complex of computer-aided design facilities (ADFC) is the whole of the different kinds of its supplements (Fig. 1.2).

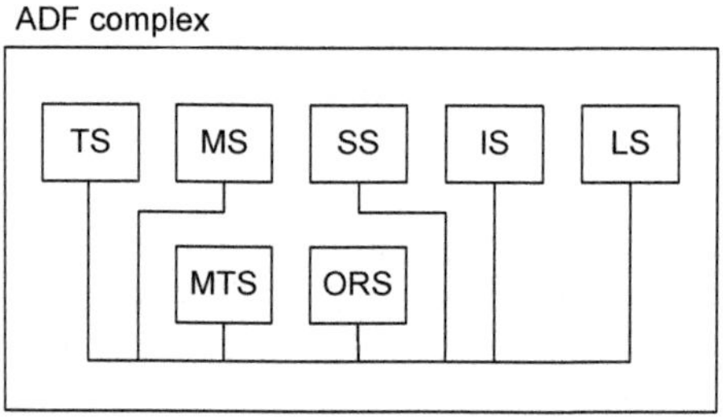

Fig. 1.2. Computer-aided design facilities complex (ADFC)

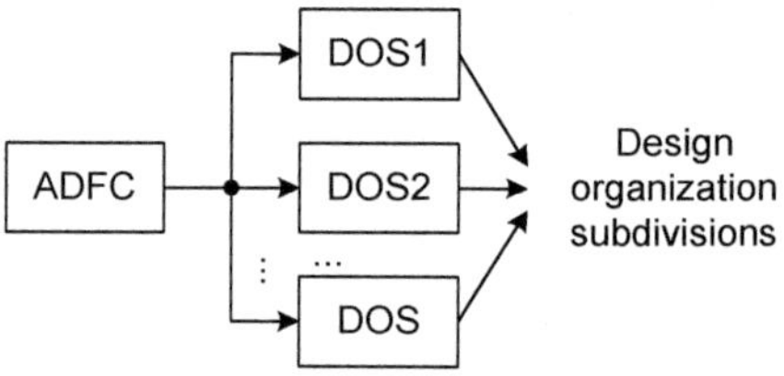

Fig. 1.3. The automated design system

Now we can formally define what a CAD system is (Fig. 1.3).

Computer aided design system is a set of automated design facilities, interconnected with the respective subdivisions of the design organization or the group of experts (system user).

In order to fulfill the objectives of the efficient EE CAD system development it is necessary to ensure:

- automation of the processes of information search, processing and output;
- perfecting the design methods based on mathematical methods and computing facilities;
- use of optimization and multi-variant design;
- improvement of the aspect of project documentation and of a part of creative work of designers, owing to automation of routine operations;

- unification and standardization of the design methods;
- interaction with the CAD systems of different level and purpose.

CAD systems consist of subsystems that have all qualities of systems per se, and are developed as separate systems.

In terms of purpose, the CAD subsystems are divided into two kinds: designing and attendant. The designing systems perform design procedures and operations, while attendant systems maintain the working ability of the designing systems.

Depending on relation to the object of design, designing subsystems are classified into object-oriented and invariant. The object-oriented subsystems perform one or several design procedures or operations, defined directly by the object of design. The invariant subsystems perform the unified procedures and operations.

So, let us examine the classification of the existing CAD systems, according to the already mentioned all-Union State Standard 17021—75.

In terms of the design object type there are CAD systems for:

- machinery construction and instrument-making industries;
- construction and architecture;
- organizational systems.

In terms of the complexity of the design objects, there are CAD systems for:

- simple objects composed of up to 10^2 parts;
- objects of average complexity, composed of 10^2 to 10^3 parts;
- complex objects, composed of 10^3 to 10^4 parts;
- very complex objects, consisting of 10^4 to 10^6 parts;
- objects of very high complexity — composed of more than 10^6 parts.

In terms of the level of automation there are CAD systems of:

- the little-automated design, where the number of automated design procedures (ADP) is below 25% of all the design procedures;
- the average-automated design, where the number of ADPs is between 25% and 50% of all the design procedures;
- highly-automated design, where the number of ADPs exceeds 50% of all the design procedures;

CAD systems are classified, according to their complexity into:

- single-stage systems;
- multi-stage systems;
- complex systems.

Then, in terms of the nature of project documentation produced there are CAD systems with the output in the form of:

- text documents;
- text and grapLSI documents;

- documents on machine-oriented media (punched cards, punched tapes, magnetic tapes, disks, drums);
- documents on photo materials;
- documents on two kinds of media;
- documents on all kinds of media.

In terms of performance, there are CAD systems of:

- low performance, which produce up to 10^5 design documents a year;
- average performance, which produce 10^5 to 10^6 design documents a year;
- high performance, which produce more than 10^6 design documents a year.

According to the number of levels of technical devices the design document systems are:

- one-level, based on average or high class computers with a set of peripheral devices;
- two-level, based on average or high class computers and one or several workstations with minicomputers; and
- three-level, based on high-class computers, one or several workstations and peripheral software-controlled equipment.

Integrated CAD is a system that disposes of alternative software components and an automated design operating system, allowing for the selection of the whole of programs with reference to a given design object, or a class of design objects.

The essential criteria for selection of the algorithmic language are as follows: computer time for running the program, written in the algorithmic language symbols; convenience of different program connections; presence of means for specialized information description; presence of modern mathematical support for the selected language; popularity of the language; availability of translators for the language for a wide class of computers.

The software of the CAD systems includes specialized calculation programs used to produce technical documentations, required by the engineers. A particular feature of these programs is their unit structure. Each software unit has a defined purpose, is simply developed, flexible and reliable. Control and coordination of the use of units are secured by the presence of the controller program.

Information, used for automated design, is allocated on several levels and organized in a hierarchical system. This system is called *archive* or *databank*. The main function of the archive is information provision as a response to a user query. Information is provided in the form of diagrams, tables, etc.

A successful use of software units is possible only if there are special programs providing “flexible” tools for input, editing, storage, transmission and output of required information, and application package processing tools, meant for realization of separate design stages.

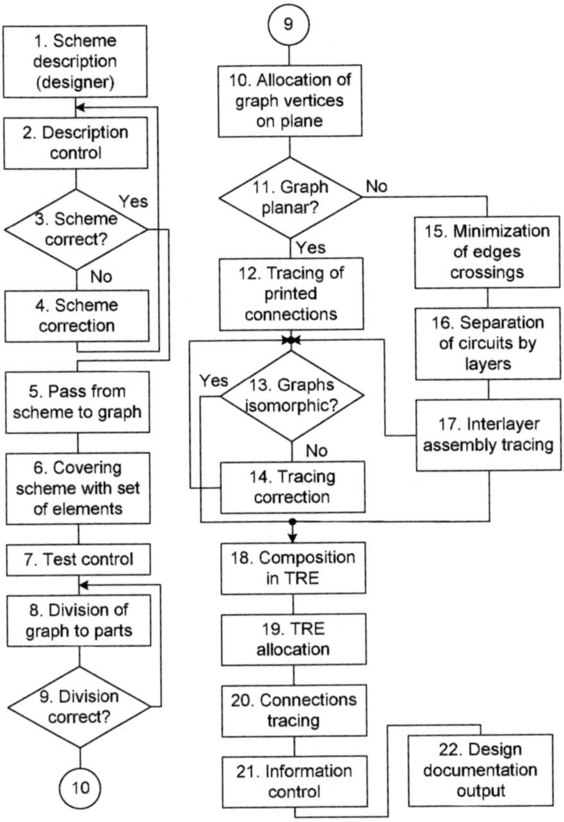

Fig. 1.4. Main stages of automated engineering design

In the process of solving the engineering design problems the following components can be nominally distinguished (Fig. 1.4): information input and control, formal scheme model development, i.e. passage from a scheme to a graph, composition, allocation, routing, and output of documentation. The composition unit usually includes the sub-units, providing for the covering of the scheme by a definite set of elements, and the algorithms of partitioning of the scheme graph into the technologically complete parts. The allocation unit is aimed at the precise, iterative and sequential allocation, with minimization of the overall circuit length and the inter-circuit crossings. The routing units include the sub-units for planarity definition, planar fragmentation of the scheme, definition of the shortest connecting trees; distribution of circuit fragments by trunks, determination of route coordinates.

The essential requirements with respect to the CAD system software units are universality, adaptability, possibility of parallel solution of several design problems, compatibility of automatic, automated and interactive modes, and database development.

Each unit in a CAD system, conform to the hierarchical principle of data processing, forms a sub-system. The source information in CAD is divided into three types: scheme description, construction description and control information. Control information sets the working modes of the software design units. The input and control unit performs the functions of information input, forming of information required for the operation of the subsequent software units, control and allocation of the information in the databank.

The most important question of EE automated design is selection of alternative variants. During automation of design, the objective is defined and written down usually as $f(x) \Rightarrow \max(\min)$, where f is some scalar function, for example, reliability of construction, ability of diagnosing the EE, etc.; and x is a vector of the controlled (variable) parameter values, for example, the number of typical elements of construction; with

$$x = \{\mathbf{x}^i\},\ 0 \le \mathbf{x}^i \le \mathbf{x}.$$

The problems of this kind are solved by finding an extreme value of the function $f(x)$ on some set X, i.e.:

$$f(\mathbf{x}) \underset{x \in X}{\Rightarrow} \max(\min).$$

In design automation, a designer faces the problem of selecting the way of operation, i.e. of finding a vector that yields the maximum (minimum) value of several functions: $f_1(\mathbf{x}), f_2(\mathbf{x}) \ldots f_v(\mathbf{x})$.

To find a solution to such problems the methods of linear convolution, control numbers, Pareto compromises, etc. are used.

A comprehensive criterion is used instead of v different criteria for the linear convolution methods:

$$F(\mathbf{x}) = \sum_{i=1}^{v} \lambda_i f_i(\mathbf{x})$$

where λ_i are positive numbers normalized in a certain way. They reflect the designer's conviction about the nature of the compromise he should accept.

Let us examine the use of some technicalities. During automated design of EE based on IC4 and IC5, the system of standards is often defined as $f^*_1, f^*_2 \ldots f^*_v$, meaning that the EE parameters should ensure $f_i(x) \Rightarrow \max$ under $f_i(x) \ge f^*_i$.

In this case, the goal function is defined as $F(\mathbf{x}) = \min\limits_i f_i(\mathbf{x}) / f_i^*$, and then the search for the value of vector $\mathbf{x}$ is performed ensuring the maximum value of $F(\mathbf{x})$. Here, $F(x) \Rightarrow \max$ means selection of such values of $\mathbf{x}$ that maximizes the ratio of the ith value of the criterion and the standard-set value. Note that the values of f^*_i are usually defined by the experts or by the designer, on the basis of experience.

Pareto compromises. In finding the solution to multi-criteria problems, rejection of the certainly "bad" solution subsets is performed. Suppose a choice of values x', and another choice, x'', so that for all criteria the inequality $f_i(x'') \geq f_i(x')$ is valid. Then, in an obvious way, the choice of x'' is preferred over that of x'. Consequently, the vectors of satisfying the inequality as x' does should be excluded from further examination. It is suggested to investigate only vectors **x** that do not satisfy the inequality for all the criteria. The set of such vectors **x** is called Pareto set.

For example, assume that the objectives of design automation are defined by two functions: $L(G) \Rightarrow \min$ and $P(G) \Rightarrow \min$, where $L(G)$ is the total number of connections in a typical replacement element (TRE), and $P(G)$ is the number of connection crossings in TRE.

Then, the feasible value of variable $x \in G$ will correspond to only one point on the plane (L, P). Identities $L = L(G)$ and $P = P(G)$ define some curve $y_1y_2y_3y_4$ (Fig. 1.5). The Pareto set consists of the region y_2y_3. Regions y_1y_2 and y_3y_4 do not belong to the Pareto set, as P and L do not increase simultaneously there.

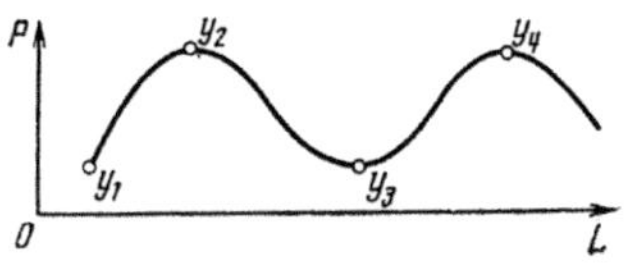

Fig. 1.5. Example of a Pareto set definition

Note that the Pareto principle does not yield a single solution. It reduces the number of variants to be examined, but the definitive selection is made by a person making decisions. In our case this person is the designer, who, helped by a computer defines the "cost" of increasing one index in terms of decrease of other indices. Using the Pareto set simplifies the decision process.

Problems, which are solved on the engineering stage, are related to multi-criteria optimization problems. Full optimization during design is in practice very difficult, because of a high number of partial criteria, variables and limitations. Currently, the most suitable optimization approach for computer-aided design is to use *sequential suboptimization methods*. First, the general optimization criterion is formed as a weighted sum of partial criteria, with weights set by the designer or determined from expert assessments. The essential point in sequential suboptimization is to find the optimum for the most important criterion, then for the less important one within the feasible set, etc. As the engineering design stages in CAD systems are performed sequentially, on every stage optimal or quasi-optimal results can be obtained, corresponding to their assigned quality factors.

During the EE engineering design the main stages are composition, allocation, routing and control. Each stage corresponds to a partial optimization criterion. For the combined calculation of partial criteria, for example, a general optimality criterion can be examined, that for engineering design problems looks like $F(G) = [K(G), R(G), T(G), G(G)]$, where G is the object of optimization (formal model of the scheme); $K(G)$, $R(G)$, $T(G)$, $G(G)$ are, in turn, general criteria for composition, allocation, routing and control stages, respectively.

Then, each of these general criteria of particular stages also consists of partial optimality criteria. As each criterion, in correspondence with the class of examined schemes, can be associated with some number that characterizes its importance in comparison with other criteria (a weight), they can be united; in other words, an additive function can be obtained

$$F(G)=\sum_{i=1}^{s}\lambda_i F_i(G)\,;\ \lambda_i \geq 0\,;\ \sum_{i=1}^{s}\lambda_i = 1\,.$$

The expression obtained allows for combination of partial criteria in a single form. For example, for engineering design problem we have $F\ (G) = \lambda_1 K\ (G) + \lambda_2 R\ (G) + \lambda_3 T\ (G) + \lambda_4 C\ (G)$. The weight coefficients λ_i are mainly based on designer's experience or expert assessments.

1.3 Review of the Modern CAD Systems

The leading tendency in the automated design of electronic devices today and in the near future is creation of the integrated intelligent CAD systems. These systems are meant to assist at all the design stages—both the complex device description on all the detailed elaboration levels (technology, device, schematics, system, architecture) and its functioning (manufacturing, behavior, scheme, system) levels. The need of developing these integrated intelligent CAD systems results, first of all, from the complexity of the objects to be designed. The wish of securing highly competitive technical and economical parameters of the DD makes it necessary to revise the limitations and the capabilities of the CAD systems, and many principles of methodology of advanced CAD systems [1].

Fulfillment of the requirements mentioned becomes necessary in the stage-wise organization of the construction process. In this case system efficiency shall in many respects depend upon the reliability and correlation of separate stages, secured by unification of input and output information, identity of carrier recording methods, computer memory distribution, etc. [2]. An important place in CAD system organization is assigned the programming language, which ought to be simple and yet assure the description of initial input information and an easy operation for the engineer. The overall decisive factor of the CAD system creation is the quantitative or qualitative benefit from automation, significantly exceeding the costs of necessary work [3]. The system developed must have high viability, meaning easy configuration, ability of changing the optimization criteria, of expanding and modifying program libraries, and of connecting to other designing systems and automated manufacturing processes [4].

Nowadays, CAD systems are being developed in two directions: on the one hand personal computer based systems with direct participation of the designer are widely used [5]. On the other hand, CAD systems based on multiprocessor systems are being developed for design without human control. It is considered that the latter direction has a better perspective as it is most automated [6]. Within both directions, though, the decisive questions are optimization algorithms, formalization of the construction tasks, information representation in computers, organization of program libraries, etc.

A CAD system should offer the following capabilities:

- automated storage of the designed device information,
- sequential expansion and perfecting of the system,
- active 'constructor-to-system' link,
- operating with optimally interchangeable construction algorithms,
- increasing system power by using multiprocessor systems and peripheral devices,
- connecting with special automatic output machines (plotters, drawing devices, etc.),
- development of construction and technical documentation [7].

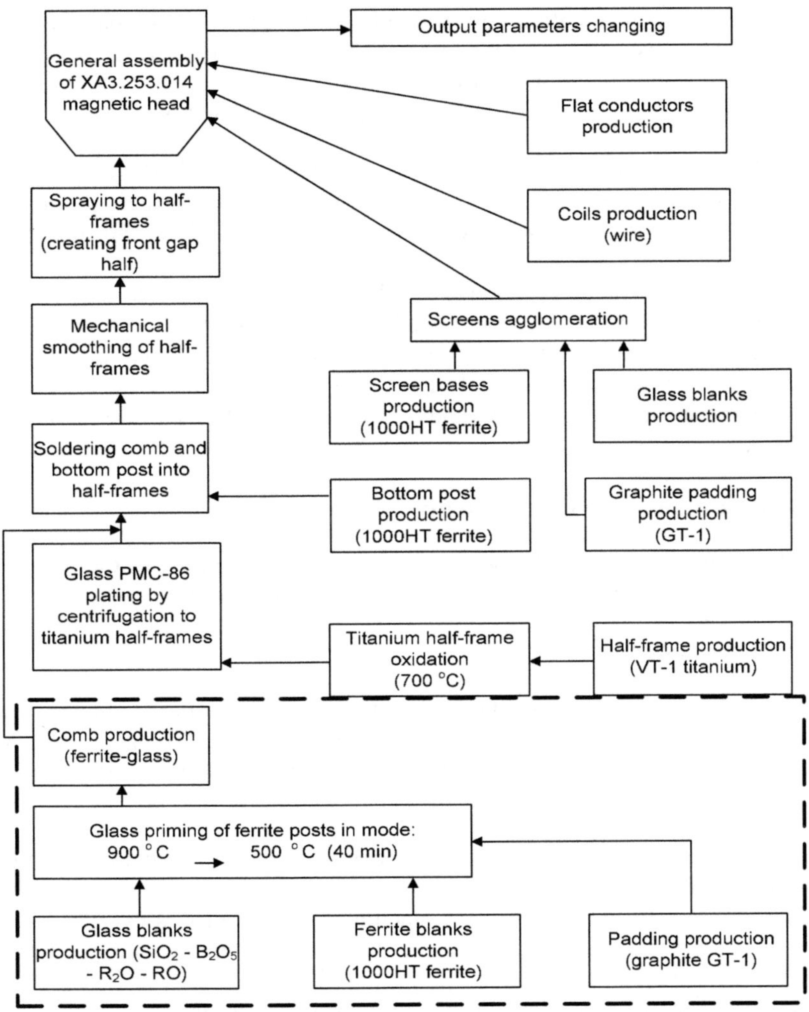

Fig. 1.6. The flow-sheet of the multi-track magnetic head manufacturing process

The quality of a CAD system is determined not only by the capability of using the system to design a wide class of devices without essential changes, but also by the degree of optimality of the algorithms and the method of information representation [8].

To create an efficient CAD system it is necessary to consider the following essential points:

- implementing automation of the search, data processing and output processes;
- design perfection based on the use of mathematical methods and computer techniques;
- use of optimization and multi-variant designing;
- creation of common data banks containing essential information;
- improvement of documentation design quality and increase of the part of constructors' creative work due to automation of the non-creative works;
- unification and standardization of design methods;
- interaction with CAD systems of different level and functionality.

Main requirements with respect to software modules of the CAD systems are: universality, flexibility, ability of parallel solving of several design problems, compatibility of automatic, automated and interactive modes, and database development [9].

The source data in the CAD systems are divided into three types: scheme description, design description and control information. Control information determines the operating modes of the design software modules. The input and control module ensures input of information, forming of the information required for the subsequent software modules, control and allocation of the formed information in the data bank.

1.3.1 Classification of the CAD Systems

The CAD systems are usually divided into three groups according to the scale of problem solving capacities: full-scale, middle-level and low-level systems.

1. The full-scale systems

The full-scale CAD systems have the broadest capabilities. Some of the systems most widespread in Russia are: Pro/Engineer, Unigraphics, CADDS 5, and CATIA. They include a large number of different functionality modules [10]. The list of typical full-scale system modules includes:

- a graphic kernel for creation of geometric models of separate components, units, and of the complete device;
- a module for creating and operating the machine-assembly processes;
- the modules for engineering analysis using the finite-element method, modeling of kinetic and dynamic mechanisms;
- modules for development of the control and life support systems;
- a set of modules for technological preparation of manufacturing process;
- models for implementing the project data control;

- a built-in database management system;
- modules for preparation and edition of the project and construction documentation.

The basic set of modules is usually complemented with different auxiliary units.

2. The middle-level systems

The middle-level systems are composed of a relatively broad set of modules, developed by the software package vendor. These systems provide high design functionality, use geometrical modeling applications with parametrical modeling. Some of them include sets of modules for the project data and assembly control. Some of such systems are: Cimatron, KONSYS 2000, Pro/Junior, and MicroStation [11].

3. The low-level systems

The low-level systems usually have a limited set of modules and include, in addition to the basic drawing automation tools, a graphic modeling tool with 3D-surface graphics, 3D body visualization unit, program generation unit for CNC equipment, and so on. They have extremely limited capabilities for parametrical design; there are no data controls, functional analysis nor machine-assembly control units. The typical systems of this class are: AutoCAD, CADDY, T-Flex, 'Kompas', 'Sprut' and others [12, 13].

1.3.2 Universal CAD Systems

The universal design systems are composed of 3 subsystems: CAD (computer aided design) units — for geometrical modeling and computer graphics, CAM (computer aided manufacturing) units — for technological pre-production, and CAE (computer aided engineering) units — for engineering calculations and analysis for verifying project decisions. A modern CAD system can provide automated support for the work of engineers and specialists at all stages of a new product design and manufacturing [14].

At the same time, the use of informational technologies in industry allows for applying them quickly and with minimal waste, taking into account the experience and the mistakes of other companies. The enterprise-wide (or full-scale) CAD systems should take a special place in the process of industrial computerization. They are an instrumental basis for all other manufacturing and organizational automation systems and, as a rule, are based on the latest achievements in automation of engineering work and manufacturing organization.

The existing CAD systems can be divided by the range of problems they solve into universal and specialized, the latter ones used either separately or as parts of universal systems. All of the universal CAD systems contain three categories of program packages:

1. Program packages for the system graphics kernel. The typical specialized programs of this category are ACIS (Spatial Technology) and Concept Modeler

(Wisdom), which realize solid-state differential geometry for geometrical model creation.

2. Packages for thorough analysis and estimation of functional and experimental qualities using modeling methods on different levels of physical representation of objects. They allow for an almost complete avoidance of the expensive production of prototypes of devices designed and of their real-life testing. Such systems are often highly complex and costly and cover a wide area of the technical modeling problems. Most widespread in this context are systems for distributed modeling using the finite-element method. Depending on the type of the device designed, manufacturing technology and operating conditions they are also divided into universal and specialized. Some of the most well-known universal systems are NASTRAN, NISA II, PATRAN, ANSYS; popular specialized CAE systems are SIMTEC and MAGMAsoft packages, Mold-Flow and others.

3. Systems for preparation of CNC machines and process environment control data. They usually have a built-in graphical editor, allowing for creation of a geometrical model of a component from its draft, used later in generation of the CNC systems control data. There are plenty of such programs, most well-known of them are SmartCAM, CIM CAD, Cimplex, PEPS, DUCT, 'Sprut', etc. They are often provided as specialized packages for certain machining modes [15].

1.3.3 Main Functional Capabilities of Advanced CAD Systems

1. ANSYS (product of ANSYS Inc., USA)

ANSYS is a multi-purpose finite-element analytic package for a wide area of engineering disciplines (static analysis, thermodynamics, gas and liquid dynamics and electromagnetism) [16]. ANSYS is a universal, so-called 'heavy' finite-element package for solving different tasks—strength, heat, electromagnetism, flow dynamics, combined multi-disciplinary analysis, optimization based on all types of analysis listed above—in a single environment (and with a single finite-element model).

The preprocessor of ANSYS allows for creation of geometrical models with the built-in software modules and for importing the available ones. The geometrical model can then be modified in any way as imported data are translated into ANSYS format and the component is not replaced with the 'read-only' finite-element grid. A user can delete the insignificant small parts; complement certain components, make grid denser or sparser, and do other important things. Sometimes the appropriate solution can be completely incorrect or cannot be found at all without these additional operations. The surface building, solid-state and skeleton geometry, as well as change applications are ensured by the geometric modeler.

ANSYS consists of the following main modules:

- geometric modeler—environment for building geometric models of objects, equipped with a set of special tools;
- grid generator—tool environment, allowing for placing the grid on 2D and 3D images;

- the load and boundary conditions unit—a tool attached to the grid built by the grid generator;
- the preprocessor visualization unit—the unit displaying the 2D and 3D images of the objects modeled.

The analytic capabilities of ANSYS

ANSYS has a lot of analytic modules allow for the analysis and development of conclusions about the quality of configurations developed. The essential ones among them are:

- linear strength: linear elasticity, sub-divisions (super-elements);
- nonlinearity: geometric (large movements, deformation and others), physical, hyper- and multi-linear elasticity, bi-polar and nonlinear kinetic strengthening, bi-polar and non-linear isotropic strengthening, anisotropic strengthening;
- dynamic analysis (frequency domain): modal (cyclical symmetry calculation, analysis of overloaded constructions), harmonic, spectral, analysis of random vibration;
- dynamic analysis (time domain): transient processes: linear, nonlinear;
- stability: linear, nonlinear;
- fracture mechanics: 2D and 3D presentation, linear, nonlinear, thermal, composition problems, reinforced materials, fragile and plastic destruction;
- thermal analysis: stationary, non-stationary, heat conductivity, radiation, convection, phase transitions;
- special models: Power Law (polymers, blood), Carreau (for general use), Bingham (suspensions), user model, multi-component flows, filtration and distributed sources, liquids with free surface (2-D VOF);
- electromagnetic analysis: magnetostatics, low-frequency harmonic analysis, electrostatics, electric conductivity, high-frequency modal and harmonic analysis;
- optimization: parametric, topological, multiple criteria, possibility of embedding user algorithms;
- fatigue analysis: fracture criterion of maximum deformation, fracture criterion of maximum tension, user defined fracture criteria, data input: direct and from the of results of analysis provided;
- solvers: direct, iterative, explicit (ANSYS/LS-DYNA).

2. Pro/ENGINEER (product of Solver Engineering Company, USA)

For creation of high-quality products, developers must have the possibility of testing products in actual operating conditions [17]. Experiments with prototypes are expensive and lengthy. In order to obtain precise results with traditional numerical analysis, highly skilled engineers with good experience are required. The Pro/ENGINEER package for engineering analysis is meant not only for use by the specialists. Designers do also have the possibility of testing the properties of parts on the initial stages of design, of optimizing their parameters, and so use the help of specialists only for solving the most complex and specific problems.

The analytic modules of Pro/ENGINEER include tools for model creation and for direct (without translation) use of the Pro/ENGINEER models. They enable import of models from other CAD systems. Pro/ENGINEER consists of the following main modules [18, 19]:

- Plastic Advisor Extension (testing of press mold design for plastic casting)—this module allows for viewing the process of press mold casting on a monitor and for estimating effectively the workability index of the detail and press mold.
- Pro/MECHANICA Structural Simulation Package (component and structure strength analysis)—a module for estimation and optimization of structural parameters of products.
- Pro/MECHANICA Motion Simulation Package (machine and mechanism kinetic and dynamic analysis)—making it possible to imitate, estimate and optimize the mechanical movement in operating conditions, and in this way to obtain the required technical parameters of the product.
- Pro/MECHANICA Thermal Simulation Package (thermal analysis of components and assemblies)—enabling to obtain the required thermal parameters of the product by analyzing the behavior of parts and structures under heat load.
- Pro/MECHANICA Fatigue Advisor (analysis of product behavior under load (fatigue analysis))—module for fatigue resistance estimation of structures subject to loads that vary periodically.
- Ce/Tol Option (calculation of dimensional chains)—the module for calculation, analysis and optimization of maximum deviation of structures dimensions.

In addition to the two above systems, the authors have also analyzed other products from the domain under consideration. These were Star-CD, LS-DYNA, ANSYS/LS-DYNA (software products of ANSYS Inc., USA), 'KOMPAS' software system (Russian Federation), MoldFlow Part ADVISER (MPA) (republic of Moldova), 'POLYGON' CAD-system (Russian Federation) [20].

The analysis of respective references showed that the main problem in design of magnetic heads (MH) for data storage devices (DSD) is the absence of automated approach and optimization algorithms that would allow for increasing the speed and quality of developing MH. Design of MH for DSD is a high-tech, expensive production stage, related to a lengthy process of selection of components with given qualities, which would meet the requirements for MH constructions. The existing CAD systems that might be effective in solving these problems are the universal packages of modules for different purposes and subsystems of design automation. The examined CAD systems allow for solving the basic problems of design of any electronic device. The specialized questions of MH for DSD design remain, though, the essential difficulty, namely: selection of materials with given qualities and their alloys, estimation of optimal MH parameters, and other important problems, which cannot be solved by the CAD systems examined.

Such problems could be solved through creation of a DSD MH automated design system, which would be integrated into the universal CAD systems here examined. This is related to the fact that almost all the of design stages of the MH are described using mathematical models of MH design and manufacturing processes, and these models are the basis of the CAD systems. Creation of mathematical models makes it possible to apply optimization algorithms for determining optimum parameter values, so as to obtain the necessary quality of the MH components. These variable parameters can be: head tip shape, front gap width, and MH material composition, magnetic and thermal fields. So, considering the complex physical processes in the MH, it is necessary to develop a design automation system to coordinate all the variable parameters with the technical parameters like data carrier velocity, data density, data transfer rate, gap between carrier and head, carrier material and thickness, recording current, required playback signal output level, and the parameters of recording and playback schemes.

1.4 Problem Statement of the DSD Magnetic Head Design Automation

Magnetic microelectronics is dynamically developing worldwide. It is based on integral planar technology of producing the micron-sized elements on magnetic films. In design of the data storage device (DSD) elements the developer faces the problem of large number of complex technical problems that must be solved in a limited period of time. Given the complexity of the developed DSD, mathematical modeling methods are widely used; they allow for formalizing the processes of taking new technical decisions and for performing the respective data analysis. As a consequence, the necessity arises of developing the corresponding automated design tools; it is oriented, first of all, at combination with the elements of artificial intelligence at all levels of the CAD system.

The DSD are used for long-time storage of large data arrays. These devices are parts of the modern computing, broadcasting, communication, measurement and other data-processing systems. Their manufacturing is, therefore, one of the leading branches of the electronic industry.

The entire variety of DSD is traditionally divided into three main areas of application, determining the type of device and the principles of its construction. Figure 1.7 shows the classification of different types of DSD.

The biggest part of the manufactured DSD belong to *Digital Recording Equipment* (DRE), used in computer equipment. Hard disk drives (HDD), magnetic tapes (MTD), optical disks (ODD) and magneto-optical disks (MODD) are used in the external memory devices of computers.

The second largest domain of DSD applications is *Audio Recording Equipment* (ARE), used in broadcasting (studio and reporting), movie technology (for synchronous recording), in private life, as well as in different industrial and transport systems (recording of talks, commands, etc.).

Video Recording Equipment (VRE) is used in telecasting, in management and control of production processes, investigation of natural resources, meteorology, private life, and also for recording of broadband signals.

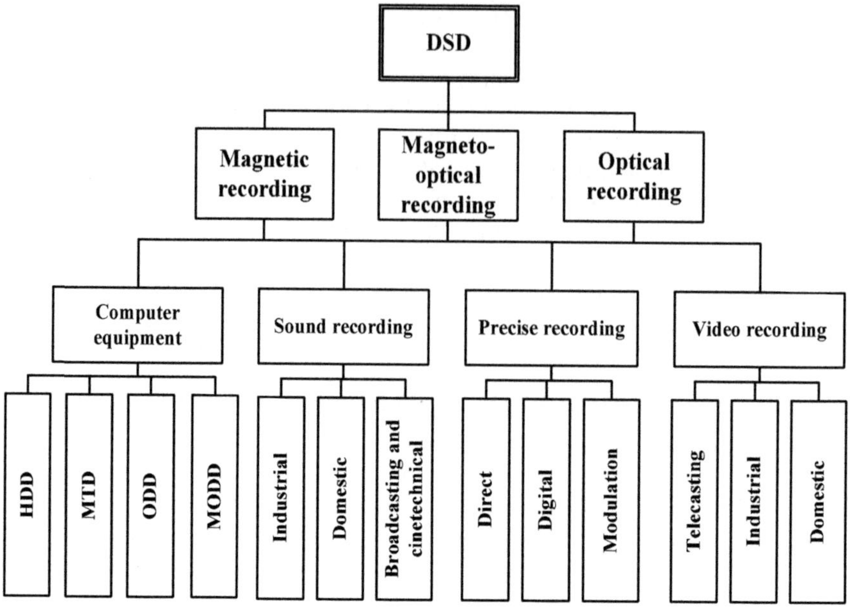

Fig. 1.7. DSD classification

Precise Recording Equipment (PRE) is used in information-&-measuring systems of aerospace equipment, transport, geology, medicine, nuclear physics, oceanography, for lathe control, and in other areas of science and technology.

The most important and scientifically complicated element of DSD, belonging to magnetic recording device (MRD), is the recording-playback head, because its quality determines the volume of information per unit of carrier area, recording and playback speed, carrier resistance against external damages, and other important characteristics of the DSD.

The process of computer modeling of MH (Fig. 1.8) can be subdivided into three main stages: model development, computer implementation of the model, obtaining the results of modeling and analyzing them.

The first stage is the preparatory one and it includes the statement of the modeling problem and of requirements with respect to the model, definition of its structural scheme and the required evaluation of modeling results, as well as a choice of mathematical description of the elements and processes of the MH. The source data for modeling are obtained on the basis of a priori information on the MH. The first stage takes about 20% of overall cost in terms of time and material [21]. The tasks of modeling are reduced to the typical ones: the definition of basic characteristics and the influence exerted on them by the MH parameters and recording mode through the investigations before the MRD design; the choice of optimum structure of MRD during the design process; the engineering calculations

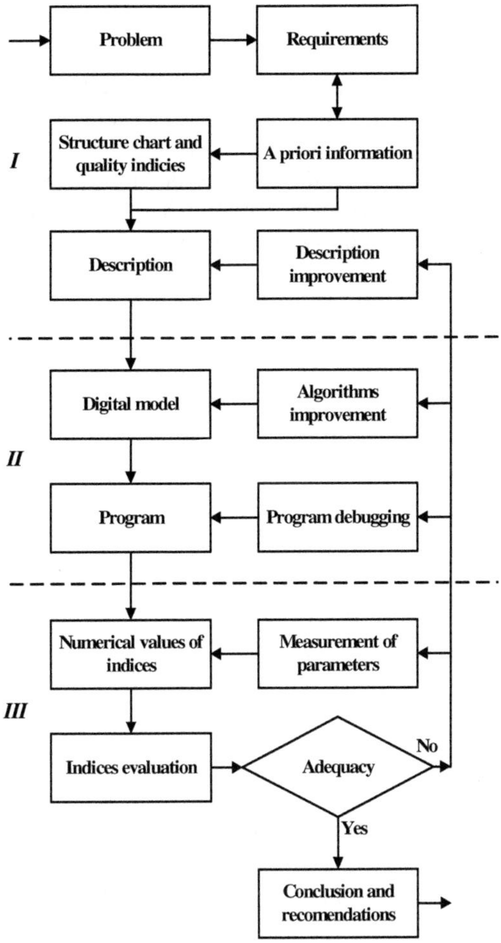

Fig. 1.8. The MH modeling process

of the MH or acquisition of information complementing the test results for estimation of MH effectiveness from the already developed devices.

The existing methods of MH development are not sufficient for realization of the whole cycle of automated design. For an illustration we showed the structural diagram of designing and manufacturing of XA3.253.014 multi-track magnetic head (Fig. 1.6).

For automation of design with respect to the modules in the dotted-line rectangle the tools are required that allow for optimizing and speeding up MH development: selection of materials with desired qualities and their composites, definition of optimal characteristics of MH, and other important problems. Hence, for the purpose of finding proper instruments for the MH design automation it is necessary to carry out the analysis of the modern CAD systems.

1.5 Conclusions

1. A classification and a review of the modern CAD systems are provided.
2. The main stages of CAD are given. The complex of CAD facilities is presented.
3. The conclusion is reached that the existing CAD systems do not allow for solving all of the specific problems of MH design automation. The necessity of developing a system for DSD MH automated design is established.

Chapter 2
Evolutionary Models of Decision Making

2.1 Evolutionary Methods and Genetic Algorithms in Artificial Systems

From among the numerous theories of evolutionary development five models of evolution will be used and considered by the authors.

The model of evolution, proposed by Charles Darwin, provides a structure for the process, in which the individuals of some population, endowed with some higher functional properties, have higher chances of reproduction, than the "weak" individuals. Such mechanism is named frequently the "survivals of the strongest". The driving forces of evolution according to Darwin are [22, 23]:

- variability, meaning the variety of organisms of each population, both due to and with respect to the hereditary features;
- struggle for existence, through which the less adapted organisms are eliminated from replication;
- natural selection, i.e. survival of the more adapted individuals, due to which the useful (hereditary) variations are collected and summarized, and the new adaptations result therefrom.

There is a spectrum of ways of development, which are all conform to the principles of Darwinian evolution. The probable ways of development define the contingency. Thus, Darwinian evolution would consist of the following points [22]:

- in nature, everything is subject to uncertain hereditary variability; the posterity appears in varying forms according to different attributes;
- organisms in nature produce multiple copies in a geometrical progression, but the number of all organisms remains on the average more or less constant;
- the principle that "the strongest survive" represents the basis for selection.

Simulation of evolution can provide algorithmic means for optimization problems on graphs. In general, evolution can be described as a multistage iterative process, consisting of random changes and subsequent selection. Thus, there is an interrelation between the definition of evolution and optimization algorithms [24]. A simplified scheme of the model of evolution, as proposed by Charles Darwin is shown in Fig. 2.1.

The model of evolution, proposed by Jean-Baptiste Lamarck, is different. It is based on the assumption that descendants inherit the characteristics acquired by an individual during lifetime. The changes in the characteristics, as asserted by

V.M. Kureychik et al.: Genetic Algorithms for Applied CAD Problems, SCI 212, pp. 23–114.
springerlink.com

Lamarck, are caused by the direct influence of the environment, bodily exercises and inheritance of the attributes, again acquired during the lifetime of the ancestors [25]. He explained one of the features of evolution of the organic world by adaptability. He explained progressive evolution and occurrence of forms, increasingly complex and perfect, by "the law of gradation"—aspiration of living beings to increase the complexity of structures. According to Lamarck, species evolve, adapting and becoming complicated, because they have the tendency to adapt and to become complicated. The reasons of the directional changes are explained differently, but they can be reduced to two:

- the directed influence of the environment;
- ability of the organism to change accordingly.

According to Lamarck, living organisms are capable of finding proper decisions to improve themselves, and are capable of carrying out these decisions. The directed variability in Lamarck's evolution is not the reason, but always the result of the evolutionary process. Lamarck's evolution corresponds to the concept of artificial evolution, as applied in science and engineering. The present authors suggest using some principles of this type of evolution in decision optimization problems. This model appears effective when a population converges in the area of a local optimum. Fig. 2.2 shows a simplified scheme of the model of evolution by Lamarck.

Then, there is the model of evolution proposed by de Frieze. Its basis resides in the occurrence of the social and geographical events, resulting in the sharp change of the situation of species and populations. Evolution, therefore, is a sequence of jumps in the development of a population, without preliminary accumulation of quantitative changes in evolutionary processes. Sometimes such mechanism of evolution is named the evolution of "catastrophes" [23]. It works through the events occurring, roughly, once in several thousand generations. The basic idea would consist in introduction of a global change into the pool of genes at the moment of accident. Fig. 2.3 shows a simplified scheme of the model of evolution after de Frieze.

Karl Popper's model of evolution is a structure implementing a hierarchical system of flexible mechanisms of management, in which mutation is interpreted as a method of random tests and errors, and selection as one of methods of management.

This method of management leads, with the help of errors, to effective elimination in the interaction with the environment. Karl Popper's evolution is formulated as twelve theses, the fundamental ones being the following [26]:

- the problems of evolution are always solved by means of a trial and error procedure;
- elimination of errors can be carried out either by full elimination of unsuccessful forms, or as evolution of management mechanisms;
- a population uses the mechanism of management, which was developed during the evolution of its own kind;

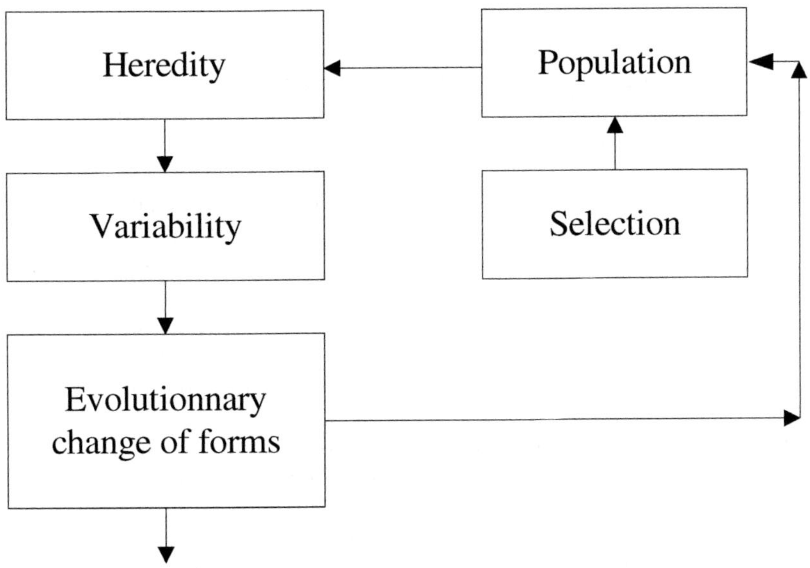

Fig. 2.1. A simplified scheme of the evolution model proposed by Charles Darwin

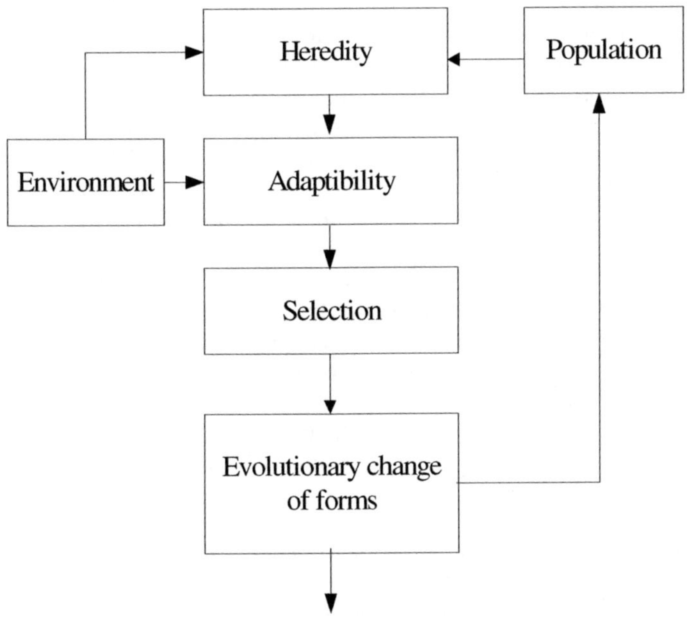

Fig. 2.2. A simplified model of the model of evolution of Jean-Baptiste Lamarck

- a population goes through the trial-and-error decision process during evolution, choosing an environment and reforming it;
- the evolutionary sequence of events is represented as $F_1 \rightarrow TS \rightarrow EE \rightarrow F_2$, where F_1—is the initial problem, TS—is the trial decision event, EE - is elimination of errors, and F_2—is a new problem.

Contrary to Darwin's evolution, where there is just one problem, the "survival of the strongest", in Popper's evolution there are also other problems: reproduction, disposal of superfluous posterity, etc. According to [26], natural systems probe the environment and actively receive information from it. The choice of the best individual in Karl Popper's evolution can be understood as the process of selection, and selection from a set of random events is not necessarily random. A simplified scheme of the model of evolution according to Karl Popper is presented in Fig. 2.4.

M. Kimura [27] proposed a model of neutral evolution with neutral selection. The idea of such selection is as follows. Let the population consist of "big" and "small" individuals. Then, evolution consists in realization of sequences of generations. The process of realization of a generation will consist of two steps.

On the first step all individuals are duplicated: big individuals have two big descendants, small ones - two small descendants. On the second step half of the individuals from the population are left in a random way with equal probability for choosing the big and small ones. It is shown in [27] that this process always converges to one of the absorbing states: all individuals are either big or small. For the populations with the number of individuals $N \geq 1000$ the characteristic number of generations T^N, required for convergence to any of the absorbing states, is equal 2^N. This evolutionary process is neutral, but as the result of evolution only one kind of individuals is selected. A simplified diagram of the model of neutral evolution is shown in Fig. 2.5. Let us remind that the block of random reduction allows for deleting exactly half of individuals from a population with equal probability.

The model of synthetic evolution, described by N. Dubinin, represents an integration of various models of evolutions, including the ones proposed by Darwin, Lamarck and de Frieze [23]. Its key feature is the recognition of stochastic nature of the mutation processes. Environmental conditions are not only treated as factors for eliminating the non adapted individuals from a population, but also as forming features of the model. A simplified, modified scheme of the synthetic model of evolution is shown in Fig. 2.6. According to N. Dubinin [23] the main points of the synthetic theory of evolution are the following:

- evolution is impossible without adaptation of organisms to conditions of the environment; the factor, forming the fitness of a structure and the functions of organisms is natural selection, which uses random mutation and recombination processes;
- natural selection, based on processes of transformation of population genetics, leads to complex genetic settings; their modifications are fixed by the stabilizing selection;

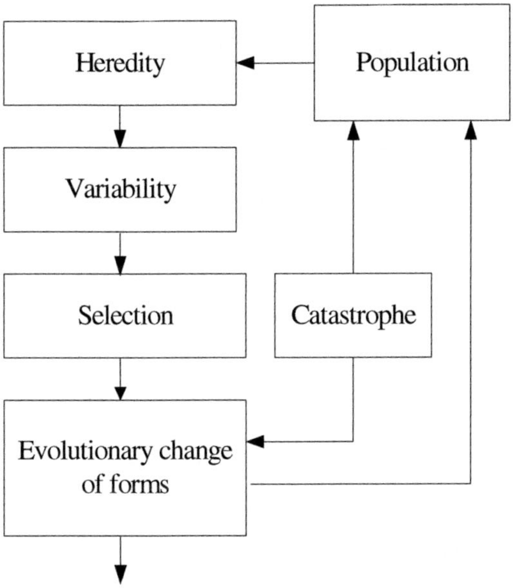

Fig. 2.3. A simplified scheme of the model of evolution after de Frieze

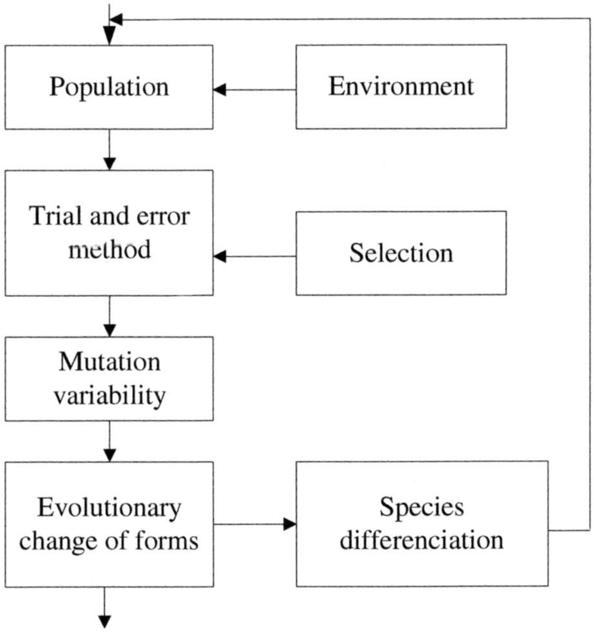

Fig. 2.4. A simplified scheme of the model of evolution after Karl Popper

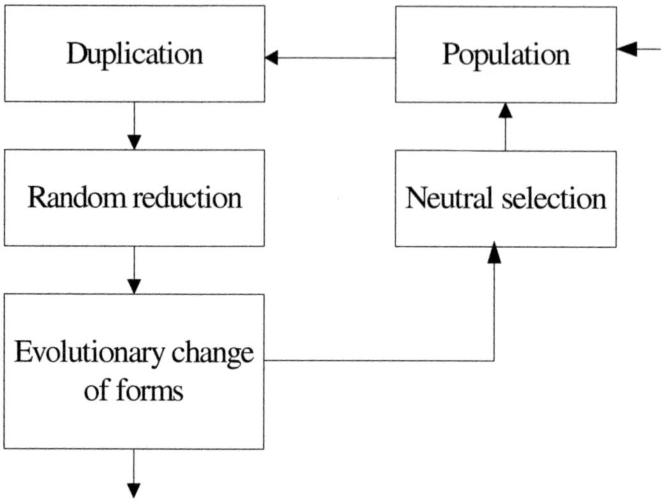

Fig. 2.5. A simplified scheme of the model of neutral evolution

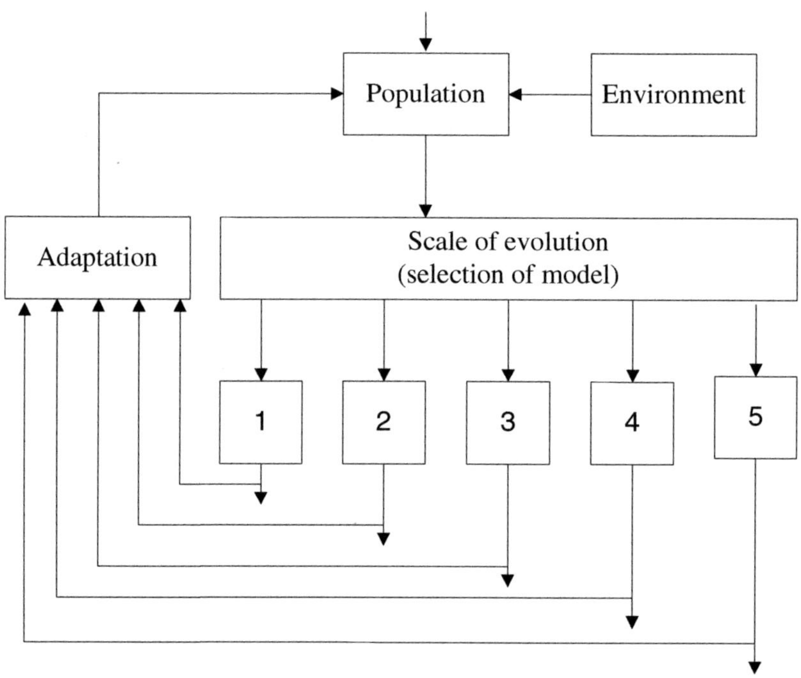

Fig. 2.6. A simplified modified scheme of the synthetic model of evolution

- in populations, hereditary variability has a mass character; occurrence of special mutations is peculiar to only isolated individuals;
- the most adapted individuals leave a lot of descendants;
- special types of evolution proceed through neutral mutations on the basis of stochastic process;
- the integrated genetic systems represent the real terrain of evolution;
- species appear by means of population evolution;
- the contradictions between the random character of hereditary variability and the requirements of selection define the uniqueness of specific genetic systems and specific phenotypes.

In this connection the present authors consider as very important the association of all kinds and models of evolutions in the modified synthetic model of evolution. Let us note that blocks 1-5 of Fig. 2.6 represent the schemes of evolution models of, respectively, Darwin, Lamarck, de Frieze, Popper and Kimura. The basic stage in each model of evolution is the analysis of a population, its transformation and the evolutionary change of forms. This is carried out on the basis of genetic search with the use of genetic and evolutionary algorithms.

Genetic algorithms (GA) represent a new area of research, which has appeared due to the work of D. Holland, his associates and students [28-35]. The GA were first applied to such problems as pattern recognition and optimization. GA as described by Holland borrow much from the terminology of natural genetics. GA represent an adaptive search method, based on selection of the best elements in a population, similarly to Darwin's theory of evolution.

The model of biological evolution and the methods of random search are the basis for the emergence of the GA. It is noted in [36] that random search (RS) appeared as a realization of the elementary model of evolution, when random mutations were modeled by random steps of the process of search for the optimum decision, and selection was "elimination" of unsuccessful variants.

The evolutionary search [36] is, from the point of view of transformation of the information in artificial intelligence system (AIS), a consecutive transformation of one fuzzy set of intermediate decisions into another. This transformation can be called search algorithm or evolution algorithm.

A GA is not a simple random search. It effectively uses the information, which has been accumulated during evolution. The purpose of GA in this book will address two basic concepts:

- to explain the adaptation processes in natural systems (NS) and AIS abstractly and formally;
- to design the AIS, which contain the homeostatic, synergetic and evolutionary mechanisms, capable of solving the optimization tasks (OT) effectively.

The central theme of the search in GA is to strike the balance between efficiency and quality for in "survival of the strongest", i.e. obtaining of optimum decisions under various uncertain and fuzzy conditions [37].

Nowadays, a new paradigm of studies based on GA and its various modifications is in use. According to [28-35, 37, 38], GA differ from other optimization and search procedures by the following:

- they basically work not with parameters but with a coded set of parameters;
- they carry out a search with a population of points, instead of a unique point;
- they use the target function values, instead of its various increments for the assessment of decision quality;
- they use not deterministic, but probabilistic, rules in the analysis of the optimization problems.

A GA takes a set of natural parameters of an optimization problem and codes them as a sequence of final length in some final alphabet. The GA works until a predefined number of generations is executed, or the decision of a predefined quality is obtained at some generation, or there is a premature convergence, when a local optimum is found and GA cannot find an exit. Contrary to other methods of optimization, GA, as a rule, analyze various areas of decision space simultaneously and are more adapted to finding of new areas with the best values of the target function (TF).

Simple genetic algorithms (SGA) consist of three operators: selection (reproduction), crossing-over, and mutation [29]. All of the GA work on the basis of the initial information, which acts as a population P of alternative decisions. A population $P = \{P_1, P_2 \ldots P_{Np}\}$ is a set of elements P_i. Here, N_p – is the size of the population. Each element of this population, P_i, represents, as a rule, one or several chromosomes or individuals (the alternative ordered or unordered decisions). The chromosomes consist of genes (elements, parts of a coded decision), and positions of genes in a chromosome refer to a locus for one position, that is a gene – is a sub element (an element in a chromosome), a locus – is a position in a chromosome, while allele – is a functional value of a gene. Genes can have numerical or functional values. The population elements in GA are referred to frequently as parents. Parents are selected from a population on the basis of predefined rules, and then produce "children" (descendants). Children and parents make up a new population. In the GA, the process of realization of one iteration of the algorithm is called (a) generation.

The primary goal of genetic algorithms consists in optimization of a target function. In other words, a GA analyzes a population of the chromosomes representing a combination of elements from some set, and optimizes TF, upon assessing each chromosome. Genetic algorithms manipulate a population of chromosomes on the basis of the mechanism of natural evolution.

According to the existing studies, it is possible to state that evolutionary methods and GA allow for solving of problems, which are difficult to solve with traditional algorithms.

$P = \{P_1, P_2 \ldots P_N\}$ it is possible to count population P dynamically changing set of final number of objects (chromosomes) $P = \bigcup_{i=1}^{N} P_i$. Let us consider a set D — space of search of all possible chromosomes, the dimension being space conterminous to number of elements in N, $\forall P_i \in D$. Then, chromosomes (the alternative decisions) are vectors in space D.

The set of genetic operators (GO) is a set of models of various types of changes, with different justifications, for chromosomes in the search space, being a mapping of a region D_1 of the search space D into another one, D_2:

$$GO: D_1 \rightarrow D_2; D_1, D_2 \subseteq D.$$

In general, GA, consist of the following: a set in the search space D; a target function; genetic operators; scheme of evolution of chromosome population.

The schematic pseudo-code of the GA is as follows:

```
Procedure GA
Begin
  initialize population P and estimate its chromosomes
  repeat in a cycle
  {
   apply the cross-over operator to chromosomes in a population P
       with some probability.
   apply the mutation operator to chromosomes in population
       P with some probability.
   apply other operators … with a predefined probability
   estimate the obtained new chromosomes
   store the best chromosomes obtained
   perform «natural selection – survival of the strongest» of chromosomes
       according to their TF value
  }
  until the condition of GA termination is satisfied
End
```

Selection (the operator of reproduction (*RO*)) is the process, by means of which chromosomes (alternative decisions); having high functional value, acquire higher chances for reproduction. The ones chosen for reproduction exchange genetic material and create various descendants. The strategy of selection, which provides an opportunity of transition from one region of alternative decisions in the search space to another, is most effective. For a random choice with frequency R the formation of parental pairs does not depend on the TF value of chromosomes P_k^t (P_k^t - is a k-chromosome, t – is a genetic algorithm generation number) and is entirely defined by the population number N:

$$R = \frac{\beta}{N(N-1)}, \tag{2.1}$$

where, β is the factor of selection acceptance, depending on the conditions in the environment, taking the values of 1, 2, 3. Another way of implementing selection is connected with the use of the values of TF. In this case there are two basic strategies. In the first one the preference is given the pairs chromosomes with close TF high TF values. In the second it is given the chromosomes with strongly differing TF values.

For realization of the first strategy, two different chromosomes are selected at random with probability

$$P(RO) = \frac{\beta}{CF(P_k^t)} \Big/ \sum_{i=1}^{N} CF(P_k^t)\, , \; \mathrm{k} = \overline{1, \mathrm{N}} \, . \tag{2.2}$$

The second strategy is implemented in the following manner: one of the chromosomes is chosen fully randomly, and the second one with probability:

$$P(RO) = \frac{\beta}{CF(P_k^t)} \Big/ \sum_{i=2}^{N} CF(P_i^t) \, . \tag{2.3}$$

Thus, the reproduction operator models natural selection.

The basic function of the crossover operator (CO) is to create the chromosomes of descendants on the basis of parental chromosomes. There are a lot of CO, as their structure defines basically the efficiency of GA. Before the start of the procedure the so-called point CO, or cutting point CO, which is usually defined randomly, is determined. It defines a place in two chromosomes where they should "be cut". Then, by exchanging the elements after and/or before the CO point between two parents, it is possible to create two new descendants. The point CO is carried out in the intelligent artificial systems (AS) in three stages:

1. Two chromosomes $A = a_1, a_2 \ldots a_L$ and $B = a'_1, a'_2 \ldots a'_L$ are chosen randomly out of the current population.
2. A number k is randomly chosen out of {1, 2... L-1}, were L is the length of a chromosome, and k is the CO point (number or value, or a code of the gene, after which the chromosome is cut).
3. Two new chromosomes are formed out of A and B by means of the rearrangement of elements according to the rule

$$\boldsymbol{A}' = \boldsymbol{a}_1, \boldsymbol{a}_2, \ldots, \boldsymbol{a}_k, \boldsymbol{a}'_{k+1}, \ldots, \boldsymbol{a}'_L \, ,$$

$$\boldsymbol{B}' = \boldsymbol{a}'_1, \boldsymbol{a}'_2, \ldots, \boldsymbol{a}'_k, \boldsymbol{a}_{k+1}, \ldots, \boldsymbol{a}_L \, .$$

In optimization problems the “greedy” crossover operators of various types are being applied. They are based on the analysis of TF values of decisions after each step of the "greedy" CO. It can be performed with two and more chromosomes, and in the limit – for the entire population. Let us present a modified algorithm of the "greedy" CO:

1. For all chromosomes of a population the TF value is calculated. A predefined number of parental chromosomes are chosen, and in the random way one of chromosomes defines the cutting point of the "greedy" CO.
2. In the chosen chromosome, for an i-th gene, located to the left of the cutting point of the "greedy" CO, the cost of distance from the i-th gene to a nearby gene is defined, i.e., a partial TF. Similar cost values are defined for all other chromosomes, chosen for the "greedy" CO.
3. In the "descendant" chromosome the gene is chosen, for which the TF value is higher or lower than the above reference, depending upon the case of maximization or minimization.
4. The procedure continues until the "descendant" chromosome is constructed in its entirety. If during the realization of the "greedy" CO a cycle or an impasse is encountered in constructing the descendant, the not yet considered genes with the best values of TF are chosen.

Mutation, reflected through the mutation operator (MO), is necessary, because it prevents the loss of important genetic material. Usually the MO is a single-point operator, which randomly selects a gene in a chromosome and in a random way exchanges it with a number (line) of located genes. The mutation operator here considered consists of two stages:

1. In a chromosome $A = (a_1, a_2, a_3 \ldots a_{L-2}, a_{L-1}, a_L)$ two different positions (for example, a_2 and a_{L-1}) are defined randomly.
2. The genes corresponding to chosen positions are rearranged, and the new chromosome is formed. So, $MO(A) = A' = (a_1, a_{L-1}, a_3 \ldots a_{L-2}, a_2, a_L)$.

In addition to mutation, there is quite popular operator of inversion, used in AIS. In the inversion operator one or more points of inversion are randomly chosen, inside which elements are inverted.

The genetic operator of inversion in GA consists of the following steps:

1. A chromosome $B = b_1, b_2 \ldots b_L$ is chosen at random from the current population.
2. Two numbers, y_{11} and y_{21}, are chosen at random from the set $Y = \{0, 1, 2 \ldots L + 1\}$, and it is assumed that $y_1 = \min\{y_{11}, y_{21}\}$ and $y_2 = \max\{y_{11}, y_{21}\}$.
3. The new chromosome is formed from B by means of inversion of the segment to the right of the position y_1 and to the left of the position y_2. So, after having applied the inversion operator we obtain B_1:

$$B_1 = (\boldsymbol{b}_1, \ldots, \boldsymbol{b}_{y_1}, \boldsymbol{b}_{y_{2-1}}, \boldsymbol{b}_{y_{2-2}}, \ldots, \boldsymbol{b}_{y_{1+1}}, \boldsymbol{b}_{y_2}, \ldots, \boldsymbol{b}).$$

The translocation operator (TO) represents the combination of CO and the operator of inversion. One break in each chromosome is made in the translocation process randomly. In the descendant P'_1 the left part, before the break comes from parent P_1 and the inverted right-hand part after break comes from P_2. Then, P'_2 takes up the left part of P_2 and the inverted right hand part of P_1.

We consider that the segregation operator (SO) represents a high interest, along with its various updating operations, which are realized on a set of chromosomes.

In the realization of the operator of removal (RO) or insertion (IO), a point or points of RO or IO are defined directly or randomly, and removal or insertion of genes or their ensembles, with calculation of TF change, are made. We shall note that the operators of removal and insertion change the size of chromosomes. When applying these operators it is necessary to take care of preserving the size of chromosomes for other operations.

Formally, therefore, we shall describe GA as follows:

$$GA = (P^0, N, P^0_i, L, TF, t, G, GO, S, R, RF, P^j) \tag{2.4}$$

where $P^0 = (P^0_1, P^0_2 \ldots P^0_N)$ is an initial population; t is a stopping criterion of GA; N is an integer number being the size of the population; P^t_i is a chromosome (an alternative decision for the problem considered); L is the length of each chromosome; TF is the target function; G is the sequential number of the generation; GO are genetic operators; S is selection; R is reduction of population; RF is recombination function; P^j is the population on the j-th step of generation.

Let us consider the chances that a GA terminates its functioning. If the value of TF for the global optimum is known, it is possible to introduce the condition for termination of GA operation in terms of, say, finding the TF value, which exceeds this global optimum by not more than a predefined ε in case we minimize TF. When the value of TF corresponding to the global optimum is not known, or priority is with the GA operating time, a stop condition can be a combination of exceeding the maximum permissible calculation time with finding a satisfactory decision in terms of TF, i.e. $TF \le TF_{gl} + \varepsilon$.

For the GA it is possible to obtain an asymptotic estimate of dependence of the convergence speed V on the values of parameters. In practical cases the speed V is linked with the operation time of the algorithm, the number of generations, population size, length of a chromosome and other size parameters. The thus determined speed is called speed of experimental convergence. It is possible to define an average experimental speed of convergence for a number of GA runs as

$$V_{avg} = \frac{1}{G} \cdot \frac{t_{\max} - t_{cc}}{t_{\max} - t_{avg} + 1} N, \tag{2.5}$$

where t_{avg} is average GA operating time; G is the number of generations; N is the population size; t_{cc} is the number of generations before fulfilling the convergence condition; t_{max} is a limitation on the number of generations from above, $t_{cc} \le t_{max}$. It is obvious that in the case of non-convergence $t_{max} = t_{cc}$, the speed is equal to zero.

Therefore, realization of a GA realization can, in general, consist of an infinite number of generations.

In solving of optimization tasks (OT) on graphs, GA offer many advantages. One of them is the capacity of adaptation to changing environment. Another advantage of GA consists in the ability of fast generation of good alternative solutions [37, 38].

A modified GA, consisting of three components, is proposed. We shall refer to the first block as to the preprocessor. This is where creation of one or more of the initial populations is carried out.

The second block will consist of four stages: choice of representation of a decision (solution); development of operators of the random, directed and combined changes; definition of laws of survival of a solution; recombination.

The third block shall be referred to as the postprocessor. Here the principles of evolutionary adaptation to the environment (to the person accepting the solutions) and of self-organization are implemented. The basic ones are as follows [39]:

Principle of integrity. In genetic algorithms the value of the target function of the alternative solutions is not reduced to the sum of target functions of partial solutions.

Principle of subsidiarity. When solving OT on graphs with GA the necessity arises of using various incompatible and complementary models of evolution and genetic operators.

Principle of discrepancy. With the increase of complexity of the problem analyzed the possibility of constructing an exact model decreases.

Principle of management of uncertainty. It is necessary to introduce various kinds of uncertainty in connection with the use of genetic algorithms.

Principle of conformity. The language of description of an initial problem should correspond to the presence of the available information about it.

Principle of the variety of ways of development. Realization of GA is multivariate and takes alternating paths. There are many ways of evolution. The primary goal is to select the ways leading to the optimum decision.

Principle of unity and contrast in relation to chaos. « Chaos is not only destructive, but also constructive », i.e. chaos in the region of feasible decisions contains, out of necessity, the order determining the required decision.

Principle of compatibility and divisibility. The process of evolution has the forward, pulsing or combined character. Therefore the model of synthetic evolution should combine these characters.

Principle of hierarchy. The GA can be built from the top downwards and from below upwards.

Principle of "Ockham's razor". It is undesirable to increase the complexity of the GA architecture needlessly.

Principle of spontaneous emergence of Prigogine. The GA allow for generating spontaneously the sets of alternative decisions among which an optimum can be found.

Principle of homeostasis. The GA are designed in such a manner that no alternative solution produced leaves the feasible area. The GA operators should allow for producing real-life decisions.

In summary, we shall note that for solving optimization problems on graphs it is necessary to take into account the dependence of the quality of solutions obtained on initial data. Therefore, in realization of concrete genetic algorithms, the present authors suggest to take into account the influence of the environment and the knowledge of the problem considered.

2.2 The Non-standard Genetic Search Architectures

2.2.1 Development of the Evolutionary Adaptation Block

Let us note that the GA can be applied hierarchically, with the use of synergetic and homeostatic principles. This is first done at the level of microevolution, where main changes take place inside one chromosome. The temporal scale of such transformations is $t \approx 4G$, where G is the number of generations. Then, this is done on the level of macroevolution, when GA act inside one population. Finally, this is performed at the level of meta-evolution, when GA are realized with respect to several populations. The temporal scale of search depends on the controlling parameters and coefficients of feedback in AIS.

The evolutionary search, following also [40], is the sequential transformation of one finite set of intermediate solutions into another. It is possible to refer to such a transformation as to a search algorithm, or algorithm of evolution. The algorithm of evolution has three features:

- each new population consists of only "viable" chromosomes;
- each new population is "better" (in sense of TF) than the previous one;
- during evolution, the subsequent population depends only on the previous one.

According to [40] a natural system (NS), realizing evolution, solves an OT through random search (RS).

One of the essential questions in the GA is the question of similarity of individuals in a population and of the relation between similarity and efficiency. The alphabet, consisting of three characters, {0, 1, *}, is used for solving these questions in GA. The tag * means that "there is no value" and 0 or 1 can be used instead of it. The concept of "template" (scheme) is introduced into GA; it is the template of similarities, describing a subset of individuals (strings) in a population with coincidence on definite positions of the string. The template in GA is a subset of chromosomes, which can be obtained from one record. Generally, it is considered that a chromosome contains 2^L templates. Then, a population of N_p individuals contains between 2^L and $N_p 2^L$ templates. The templates of a definite short length are called "building blocks". The size of building blocks is very important for the speed of solution finding and quality of the solutions found. Let us consider the examples of building blocks for different templates. For example, in the template (***1) the building block will be the "1". In the template (10***) the building block will be the string "10". Let us remark that the building blocks should be chosen on the basis of knowledge of the task considered, treating them as “bricks” used to construct a “building”, that is – the solution with the best target function value. In implementing the GA the condition should be satisfied that the building blocks are broken down only in the extreme cases, indicated by the person, who accepts the solution (the decision-making person — DMP).

We shall introduce two characteristics for quantitative assessment of templates in GA: the order of the template — $o(H)$; length of the template — $\delta(H)$. The order of the template is the number of the fixed positions (in the binary alphabet this is the number of ones and zeros in the template). The length of the template is the distance between the first and last positions of zeros and ones. The value of the ratio $f_i(x) \Big/ \sum_{i=1}^{L} f_i(x)$ is referred to as the probability of selection of copies (chromosomes) in OTDM:

$$p_i(GA) = \frac{f_i(x)}{\sum_{i=1}^{L} f_i(x)}, \tag{2.6}$$

where f_i (x) is the value of TF of the i-th chromosome in a population, and $\sum_{i=1}^{L} f_i(x)$ is the total value of the TF of all the chromosomes in a population. The value (2.6) is also called the normalized value. The expected number of copies of the i-th chromosome after the operation of reproduction (RPO) in GA is defined as

$$n = p_i(GA) \cdot N, \tag{2.7}$$

where N is the number of chromosomes being analyzed. The number of the chromosome copies, passing over to the following generation, is sometimes defined on the basis of the expression

$$n = \frac{f_i(x)}{\bar{f}(x)}, \tag{2.8}$$

where $\bar{f}(x)$ is the average value of TF in the population.

Suppose there is some template H, which is present in a population P^t. We shall denote the number of the chromosomes of P^t, which correspond to template H, by m(H,t), t denoting the number of generation or a nominal time parameter. After having obtained a set of not intersecting populations of the size N_p, by moving a part of chromosomes from population P^t, we expect to obtain m(H,t+1) representatives of the scheme H in the generation of population P^{t+1}. The expression for m(H,t+1) is given by the formula:

$$m(H, t+1) = m(H, t)\, N_p f(H) / \sum_{i=1}^{L} f_i(x), \tag{2.9}$$

where f(H) is the average TF of a chromosome, represented by the template H in generation t.

If we denote the average TF for the entire population as

$$\bar{f}(x) = \sum_{i=1}^{N_p} f_i(x) / N_p,$$

then:

$$m(H, t+1) = m(H, t) \times f(H) / \bar{f}(x). \tag{2.10}$$

It is obvious from the above expression that the population corresponding to template considered grows according to the rate equal the ratio of the average TF of the template to the average TF of a population. Let TF of the template be higher than the average by $c \times \bar{f}(x)$, where $c > 0$ is some constant. Then we have:

$$m(H, t+1) = m(H, t) \times (\bar{f}(x) + c \times \bar{f}(x)) / \\ / \bar{f}(x) = (1 + c) \times m(H, t). \tag{2.11}$$

Since t = 0 and considering that $c > 0$ is constant, we obtain the equation:

$$m(H, t) = m(H, 0) \times (1 + c)^t. \tag{2.12}$$

This expression shows that the number of good chromosomes in a population grows exponentially with time, and the number of chromosomes having TF value below the population average, decreases exponentially.

The lower bound on the probability of survival of a template can be approximately determined for any population after application of CO. There are L–1 points available for the crossover inside the string of the chromosome, as the total length of a chromosome is L. As the template in GA survives, when the point of application of the CO does not get outside of the "defined length", the probability of survival for a simple CO is as follows:

$$P_1(s) = 1 - \delta(H) / (L-1). \tag{2.13}$$

If CO is implemented through random choice, with, for instance, probability P(CO), the probability of survival of a template is given as:

$$P_2(s) \geq 1 - \alpha P(CO)\, \delta(H) / (L-1), \tag{2.14}$$

where α is the coefficient defined for the application of CO, $\alpha \approx 1$–4. If we assume the independence of the recombination operator and CO, we can write down the joint estimation of the effect of these operations:

$$m(H,t+1) \geq [m(H,t) \times f(H) / \bar{f}(x)] \times [1 - \alpha P(CO) \times \delta(H) / (L-1)]. \tag{2.15}$$

It follows from the above expression that the templates of chromosomes with TF exceeding the average, and a short defined length $\delta(H)$ have the possibility of growing exponentially in successive populations.

Let us consider now the influence of the here applied MO on the chances of survival of a chromosome with the best TF to the following generations. If, after application of an MO, a chromosome with template H is to survive to the following generation, all of its specific positions should survive. The survival of a single position has the probability of (1–P(MO)). The template H survives when each of its $o(H)$ fixed positions survives. For small values of P(MO) << 1, the probability of survival of a template can be approximated by the expression:

$$P_3(s) = 1 - o(H) \times \beta P(MO), \tag{2.16}$$

where β is the coefficient defined for application of MO, $\beta \approx 1$–3. Then it is assumed that the arbitrary template H gets an expected number of copies in the following generation after the application of the recombination operator, CO and MO:

$$m(H,t+1) > [m(H,t)\, f(H) / \bar{f}(x)] \times \times [1 - \alpha P(CO)\, \delta(H) / (L-1) - o(H)\, \beta P(MO)]. \tag{2.17}$$

Let us refer to this expression as to the fundamental operational theorem of the GA [29, 30].

The probability of survival of the template H in the subsequent generations after the application of the inversion operator can be written down as:

$$P_4(s) \geq 1 - 2 \cdot \varphi P(IO) \cdot \left[\frac{\delta(H)}{L-1} - \frac{\delta(H)^2}{(L-1)^2} \right], \tag{2.18}$$

where P(IO) is the probability of the choice of a chromosome, corresponding to the template H, from the population on a given step of generation (the probability of the inversion operator); φ is the coefficient used in application of the operator of inversion, φ≈1–3. Then the fundamental theorem of GA (2.17), accounting for the operator of inversion, becomes:

$$m(H, t+1) > [m(H, t)f(H)/\bar{f}(x)][|1-\alpha P(CO)\delta(H)/(L-1)-o(H)\beta P(MO))-P_4(s)|] \quad (2.19)$$

We shall simplify this expression as follows:

$$m(H,t+1) > [m(H, t)\times f(H)/\bar{f}(x)][|P_2(s) - P_3(s) - P_4(s)|]. \quad (2.20)$$

The probability of survival of a template on the following generation upon the application of the segregation operator is defined by the formula

$$P_5(s) \geq [1-\gamma P(SO)\delta(H)/(L-1)]/N_p, \quad (2.21)$$

where γ is the coefficient which defines the application of the segregation operator, γ≈1–3.

The probability of survival of a template for the translocation operator is

$$P_6(s) \geq [1-\alpha P(CO)\delta(H)/(L-1)]\varphi P(IO). \quad (2.22)$$

The probability of survival of a template on the following generation for the use of the deletion operator is defined by the formula

$$P_8(s) \geq [1-\lambda P(OB)\delta(H)/(L-1)]/N_p, \quad (2.23)$$

where δ is the coefficient, which is used in the application of the deletion operator, δ≈1–2.

The probability of survival of a template on the following generation upon the application of the insertion operator given as

$$P_8(s) \geq [1-\lambda P(OB)\delta(H)/(L-1)]/N_p, \quad (2.24)$$

where λ is the coefficient, which is used in the application of the insertion operator, λ≈1–2.

Then the fundamental theorem of GA, considering the action of all the genetic operators in solving the engineering optimization tasks, becomes

$$m(H,t+1) > [m(H,t)f(H)/\bar{f}(x)][|P_2(s)-P_3(s)-P_4(s)-P_5(s)-P_6(s)-P_7(s)-P_8(s)|] \quad (2.25)$$

The fundamental theorem of GA shows the asymptotic number of the "surviving" schemes for every iteration of the SGA implementation. The expression (2.25) shows the asymptotic number of chromosomes "surviving" upon the implementation of the crossover, mutation, inversion, segregation, translocation deletion and insertion operators on each generation. It is obvious that this number is approximate. It varies depending on the probability of application of a definite GA procedure. The values of TF for the separate chromosome and across the entire population, as well as the knowledge of the concrete task being solved, exert a particular influence on the number "surviving" and "dying" chromosomes for an implementation of GA.

Lemma 2.1. *If on some step of generation of GA $P_1(GA)$ is the probability that a chromosome P_i produces a descendant on this step and $P_2(GA)$ is the probability that P_i is deleted during this step, the expected number of the descendants of chromosome P_i is $P_1(GA)/P_2(GA)$.*

The quantity of templates, which can be obtained from a population of size N_p with chromosome length equal L, is between 2^L and $N_p \cdot 2^L$. The number of the "effective" templates is approximately equal N_p^3.

Therefore, the number of the "surviving" schemes is proportional to the cube of the size of the population and is of the order $N(s) = \omega N_p^{\ 3}$, where ω is a definite coefficient ($\omega = 1–4$).

Let us formulate the description of GA as a scientific theory. Assume that each initial concept and relation of the axiomatic theory of expert systems (ES) is assigned a corresponding concrete mathematical object. The collection of such objects is called field of interpretation. A statement U* about the elements of the field of interpretation, which can be true or false, is assigned a corresponding statement U of the ES theory. Then, it is possible to state that a statement U of the ES theory is, respectively, true or false in the given interpretation. The field of interpretation and its properties are the object of consideration of a special GA theory, which can, in particular, be axiomatic. This method allows for proving the statements like: if the GA theory is consistent, the ES theory is as well.

Suppose that the ES theory is interpreted in terms of the GA theory in such a manner that all axioms Ai of the ES theory are interpreted as the true opinions A_i^* of the GA theory. Then, any theorem of the ES theory, that is any statement A, logically derived from the axioms $\{A_i\}$ in ES, is interpreted in GA by some statement A^*, derived in GA from the interpretations $\{A_i^*\}$ of the axioms {Ai} and is, therefore, true.

The method of interpretations allows also solving the problem of independence of the axiom systems: for the proof that the axiom A of the ES theory does not depend on other axioms of this theory, that it is not derived from them, and, therefore, is necessary for obtaining the entirety of the given theory, it is enough to construct such interpretation of ES, in which this particular axiom is false and all other axioms of this theory are true. The specific rendition of the concept of axiomatic theory is the concept of a formal system. It allows for representing mathematical theories as exact mathematical objects and for building a common theory or meta-theory of such theories. The common scheme of the arbitrary formal ES system construction is as follows:

1. The language of the ES system: the apparatus of the algebra of logic; the set theory; the graph theory.
1.1. An alphabet is a list of elementary characters of the system: binary, decimal, letter, Fibonacci, etc.
1.2. The syntactic formation rules, according to which, out of elementary characters, the formulas of the ES system are established. These rules concern:

- construction of the evolution model;
- design of populations;
- constructions of the TF;
- developments of genetic operators;
- reproduction of populations;
- recombination of populations;
- reduction.

A sequence of elementary characters is treated as a formula only in the case it can be constructed with the help of formation rules.

2. The axioms of the ES system. A set of finite formulas, called axioms of the system, are designated. There are a large number of axioms in ES. The following are the main ones:

- "survival of the strongest", i.e., the transmission of solutions with the best goal function values to the following generation;
- the size of a population after each generation remains constant;
- mandatory application in all genetic algorithms of the crossover and mutation operators.

3. The production rule of the ES. A finite collection of predicates $P_1, P_2, \ldots, P_k$ is fixed on the set of all formulas of the system. Suppose that $P(x_1 \ldots x_{ni+1})$ means that for any of these predicates ($n_i > 0$) if the statement $F_1 \ldots F_{ni+1}$ is true for the given formulas $P(F_1 \ldots F_{ni+1})$, then we can state that the formula F_{ni+1} directly follows from the formulas $F_1 \ldots F_{ni+1}$ by a rule P_i of the system.

The definition of the formal ES system as an exact mathematical object is satisfied by the definition 1.2.3. Thus, the degree of accuracy is defined by the level of accuracy of the alphabet definition, formation rules and production rules. As the output of the ES system we consider any finite sequence of formulas, in which each formula either is an axiom of the ES system, or follows directly from any prior element of this sequence of formulas through one of the production rules P_i of the system.

Any concrete mathematical theory of ES can be translated into the language of a suitable formal system in such a manner that each false or true sentence of the ES theory is expressed by some formula of the system.

There exists knowledge, enabling the construction of the approximating models for many solution making problems. This can lead to reduction in problem size and time of calculations, to simplified simulation of functions, to reduction in the number of simulation errors in the use of GA.

Let us note that the solution of the main OT of solution making is carried out by sequential, iterative, exact or combined algorithms [41]. Essentially, these algorithms implement the process of perfecting the initial candidate solution, obtained by one of the random search methods. This approach, besides known advantages, has definite weak points:

- all algorithms, except for the exact ones, frequently stop at local solutions, which can be far from optimum, and can hardly leave the local "holes";
- exact algorithms can work only with very limited numbers of units, insufficient for solving OT of solution making;
- only one candidate for solution is used as the starting point.

The proposed OT of solution making uses not just one, but several alternative initial solutions. Depending on the complexity of the processed information, the initial solutions can be associated with stochastic, deterministic or combined algorithms. The successively obtained solutions will be handled by the genetic search algorithms, adapted to the optimization tasks of solution making, being solved, with the use of synergetic and homeostatic principles (GSMS). The basic structure of the thus conceived GA can be as shown in Fig. 2.7.

Here, the GSMS (Genetic Search Management System) block allows for controlling the genetic search (GS) process on the basis of feedbacks, with consideration of the synergetic and homeostatic principles. According to this scheme, a subset of solutions P of the analyzed OT solution making is obtained in the first stage by a random, directed or combined method. These solutions will lead to production of subsequent generation or population of solutions at steps t ($t = 0, 1, \ldots, T$). Then, the values of TF are entered or calculated. Calculation of TF is a very difficult task, and the quality of the solutions obtained depends on the accuracy of TF.

Let us note that it is desirable to build a TF for every OT of solution making. It is necessary to use the knowledge of the concrete task and the elements of ES and GSMS when constructing TF [37, 38, 40]. The ranking and sorting of the populations of solutions P is performed on the basis of TF. Then, as a result of various methods of selection in P, various GO are selected and tuned for application. After the application of CO, DMO, IO, SO, TO etc., a new subset of solutions P' is produced. It is joined with the initial subset of solutions. A new set is produced, $P_{GA} = P \cup P'$. Using TF, P_{GA} is analyzed. From our point of view all elements of P_{GA}, whose TF value is worse than a given threshold, are the unpromising solutions and they are deleted from the P_{GA}. A new set, P'_{GA}, emerges. And $|P'_{GA}| = |P|$. If this condition is not satisfied, for instance $|P'_{GA}| < |P|$, then the element with the best characteristics from among the discarded ones is entered into P'_{GA}. The resulting set P'_{GA} is treated as a new current population of solutions and the process can be repeated iteratively on the basis of GSMS so as to obtain a set of or one optimal solution.

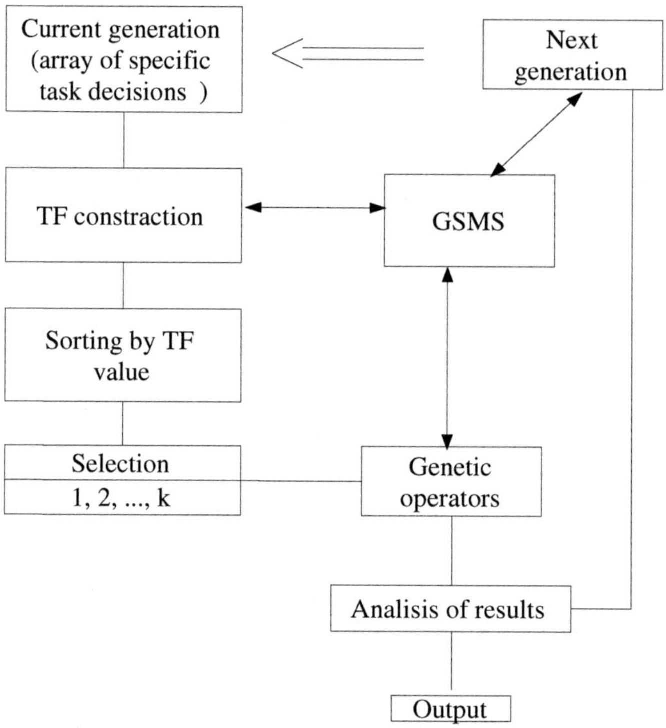

Fig. 2.7. The basic frame of GA accounting for GSMS

Let us remark that such a strategy allows for finding of locally optimal results faster. This is linked with the parallel processing of the alternative solutions set. And it is possible to concentrate search on obtaining more promising solutions in such a scheme. Note that at each iteration of GA it is possible to carry out various changes in the treatment of the more and less promising solutions. The time-wise complexity of such algorithms essentially coincides with the complexity of fast iterative algorithms and the number of key operations of these algorithms is contained between O $(K_i n)$ and O $(K_j n^3)$, where K_i, K_j — are coefficients, n – number of elements. This level of complexity leaves open the perspective of using GA with GSMS for solving OT. It should be reminded that in the described SGA frame there is a serious disadvantage consisting in the convergence of algorithms to local optima, which can be far from the global one. To eliminate this disadvantage, numerous modified GO, selection schemes and various architectures of GS are used. Yet, using such approaches does not always enable avoiding local optima [42—49].

Some new non-standard GS architectures of GS appeared recently, permitting in most cases to effectively prevent the premature convergence of algorithms. These are the methods of migration and artificial selection [50], meta-genetic

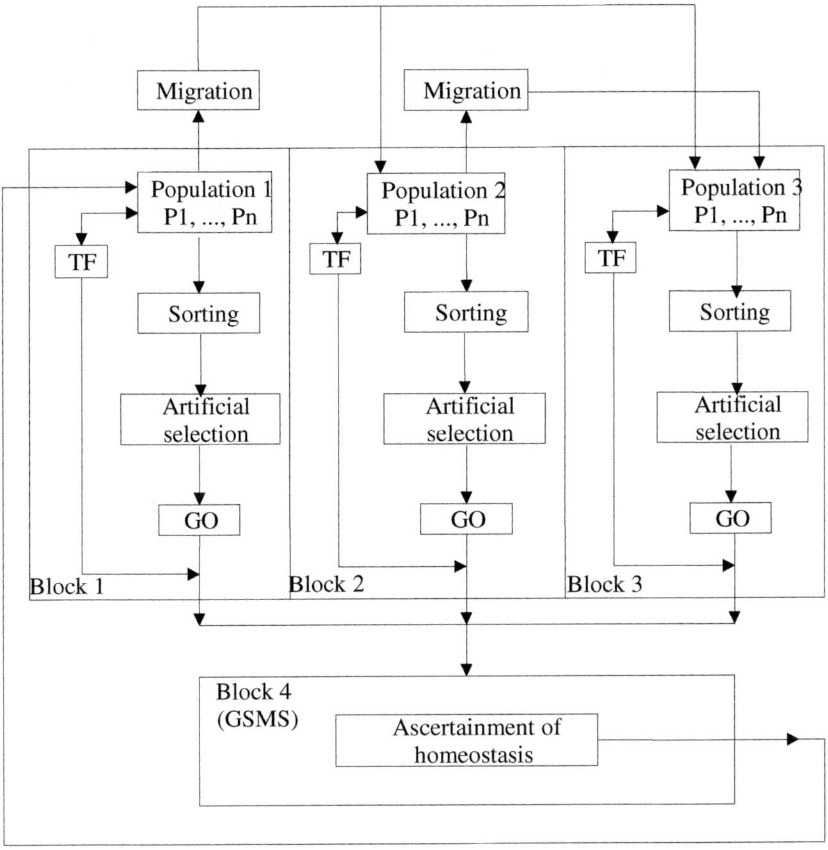

Fig. 2.8. The modified scheme of migration and artificial selection

parametric optimization [42], stochastic iterative genetic search methods [51], methods "of intermittent equilibrium" [52], associations of GS and simulated annealing (SA) [53], which were also proposed by the present authors for GS with the use of GSMS [37, 38].

In the procedures proposed by the authors [37, 38], macroevolution is realized on the basis of ordinary GA, i.e. we deal not with one population, but with a set of populations. GS is carried out in this case by the association of chromosomes from various populations. Compared to [50], a modified GS architecture with migration and artificial selection (Fig. 2.8) is proposed. Blocks 1—3 in Fig. 2.8 represent the ordinary GA. Let us note that in each block artificial selection is carried out. Selection on the basis of a "roulette wheel" is performed in the first block. Selection on the basis of a predefined scale is used in the second block. Elite selection is used in the third block. Each time, the best representative of a population is sent to the migration block. The sequential chain 1-2, 2-3 realizes the link between blocks 1 and 3. The establishment of balance on the basis of both synergetic and homeostatic principles, and handling of all GS processes is carried out in block 4.

Note that it is possible to organize various numbers of links between blocks according to the complete graph, of a star etc. Such a scheme (Fig. 2.8) can be extended to up to N of blocks in case we dispose of sufficiently ample computing resources. Additionally, N-1 blocks can carry out evolutionary adaptation in a parallel way and exchange the best representatives of solutions through migration blocks. The last block gathers the best solutions, can terminate operation or continue genetic optimization. Such a scheme of optimization, contrary to the existing ones, allows for quitting of the local optimum in many cases. Various modifications of this search scheme will be given further on.

The process of meta-genetic optimization consists in realization of the following scheme (Fig. 2.9):

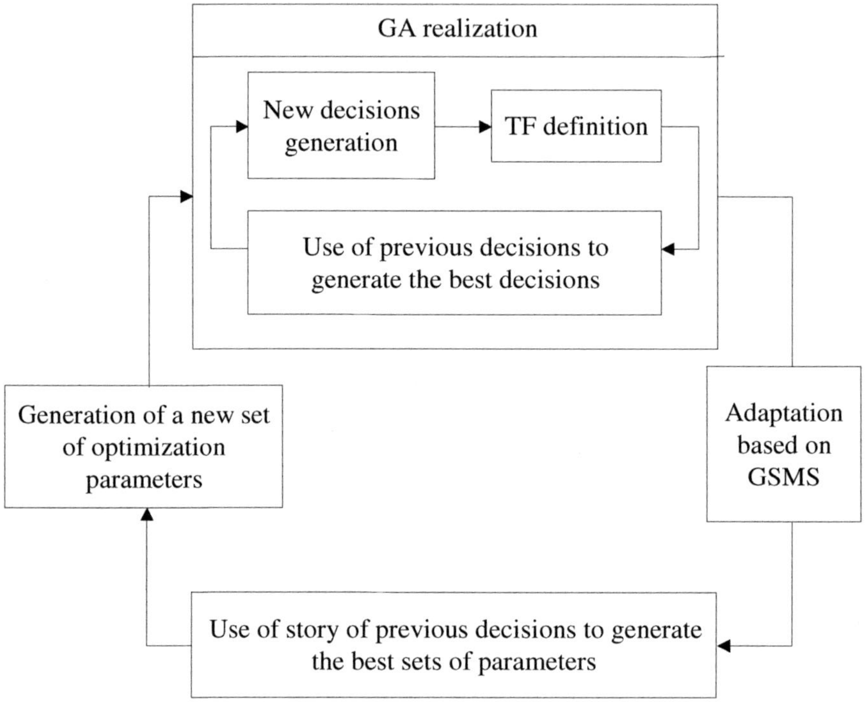

Fig. 2.9. The meta-genetic optimization process

The essential one is the first block, in which the implementation of GA, generation of new solutions, calculation of TF, and the use of previous solutions for generation of the best results, are carried out. The second block allows for the use of "history" of previous solutions for the generation of the best set of parameters. A new set of optimization parameters is generated in the third block.

Using the meta-genetic optimization process, it is possible to generate the initial populations in random, directed or accidentally directed ways, to simulate them and to realize the GA based on implementation of various GO. It is possible to select parents from the population at random with an arbitrary or predefined probability. Moreover, the probability of execution of each operator can be determined as a function of the TF value. The final set of parameters is selected after a simulation for a finite population. Let us note that it is possible to build a concrete meta-genetic algorithm for any OT of DM.

The association of GA and SA makes it possible to obtain better results at the expense of complication of the optimization procedure [54]. For example, it is possible to identify a subset of parents with the best characteristics on the basis of SGA and to apply the optimization procedure MO for one of them (the best one) or for some subset [53]. It is possible to make such an association in a variety of ways. Unfortunately, the procedures of SA require important computing expenses. It is possible to select one branch of the solution tree and to carry out SA, as the search throughout the entire tree may be too burdensome.

In [55—57] the scheme of search based on Darwinian evolution (DE) and Lamarckian evolution (LE) is described. DE is implemented as GA, into which the LE algorithm is built. It contains the Lamarck parameters (scale, G), used to control Lamarckian evolution. Scale is a real number between 0 and 1, defining the percentage of the population, to which the LE is applied, and G is the number of GA iterations. The first unit of LE realizes the case, when the user wants to apply the Lamarckian evolution to the entire population (scale = 1). This unit collects the complete set of chromosome population and the LE algorithm is executed for them. The second unit is used when the percentage of the population addressed is greater than 0. Then the number of chromosomes thus designated for the process is accordingly handled with the algorithm.

The phenotype is considered for use in the retrieval algorithm and the appropriate genotype (which is identical to phenotype in the case considered) is corrected upon reaching the best results. It is possible to include any set of the retrieval DE and LE algorithms in the scheme of the joint algorithm. GA uses the nested hybrid approach, in which the generation is defined for the analysis of the Lamarckian evolution and accumulation of respective statistics. This allows for a definite effectiveness in avoiding the premature convergence of algorithms.

Let us consider the new architectures of search. A building block, on which the multilevel AIS for solving the OT can be constructed, is shown in Fig. 2.10. Here, creation of a new population P' is not only influenced by the evolutionary adaptation block (EAB), equipped with GSMS, but also by the external environment. Such building blocks (BB), like "bricks", can be used to construct AIS of any complexity.

In Fig. 2.11 the scheme of a BB with relative balance is shown. Here, units have various properties, but, being interconnected with each other, support relative balance.

The proposed schemes of BB for development of multilevel hierarchical processes of GS help in many cases to quit local optima in solving the OT of solution making.

Note that we deal with generation of feasible alternative solutions in uncertain, blurred conditions, their assessment, and choice of the best one [58, 60]. Recently, computerized decision making support systems (DMSS) [58—60], based on

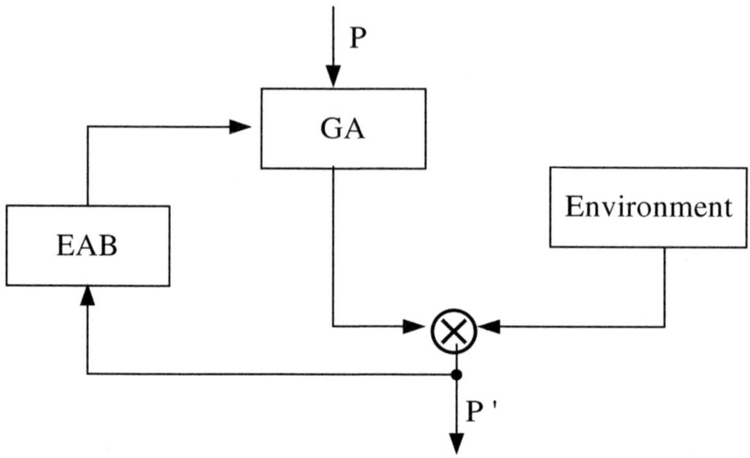

Fig. 2.10. The scheme of the building block (BB)

formalized methods of solution generation, calculation of TF, and algorithmic process of solution evaluation have become the subject of high interest.

An important duty of the DMSS is to find the order, hidden in the chaos of possible alternative solutions, surrounding us. Selection of a subset of solutions is the gist of problems of choice and decision making.

A tuple $\alpha=<N,\Theta>$ is called the task of solution making (where N is the set of variants of task solutions; Θ is the principle of optimality, providing representation of quality of variants, in the elementary case as preference over variants). A set $N_{op}\subseteq N$ is called α solution of the task, obtained due to application of the principle of optimality.

As it was already noted, it is proposed to the solution making to use the analogies of solutions to similar tasks, which have arisen through the use of NS during the entire evolutionary process. According to existing opinions (though appearing as disputable to the present authors), evolution does not carry a strictly defined internal nor external purpose. It is the effect of physical laws acting in an independent manner in a population of units [57].

The task α is solved as follows. We construct the set N, if it is possible, i.e. we define the variants, and then solve the task of choice. The standard procedure of DMSS consists of the following essential phases [61]:

- generation of the possible solution alternatives;
- evaluation of solutions (construction and realization of the TF);
- coordination of solutions, analytic presentation of the development of a situation;
- decision making (selection of a solution or a group of solutions);
- evaluation of correspondence of the accepted solutions to the envisaged purposes.

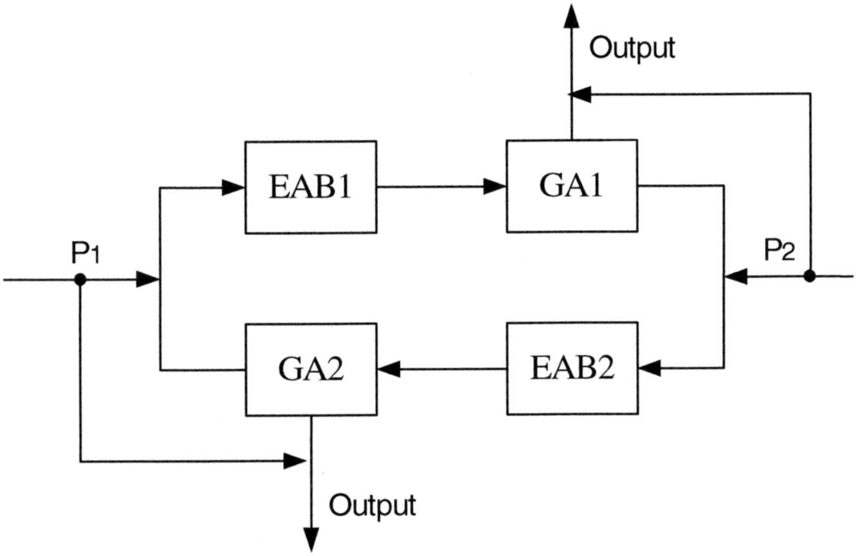

Fig. 2.11. The scheme of BB with relative balance

One can notice a lot of commonality in comparing this framework with the one of genetic or evolutionary search. In this connection the concepts, principles, architecture and algorithms of GA can be effectively used in DMSS.

Fig. 2.12 shows the architecture of solution making with genetic search using genetic search management system (GSMS). The compensator unites the properties of the evolution models. The reductor regulates the dynamically changing size of the solution population. Thus, the best chromosomes are destined for blending of populations and leaving the local optimum. The chromosomes filter block allows increasing the efficiency of evolution implementation and for speeding of the solution making.

An important question is implementation of evolution with several GA, whether the parallel, sequential, or some sort of sequential-and-parallel configuration is effective in reaching the solution to OT of solution making for the large dimension. We propose the use of Plato graphs [62—66], i.e. regular polygons, which, as held in the ancient doctrines, possess internal beauty and harmony [67, 68]. In Figs. 2.13 (a) and 2.13 (b) simplified schemes of GS, based of Plato's graphs are shown. Fig. 2.13 (a) is based on tetrahedron and Fig. 2.13 (b) - on hexahedron (cube). Fig. 2.13 presents the "one-to-many" input-output systems. In this context let us note that there can also be "one-to-one", “many-to-one", and "many-to-many" basic systems. The efficiency of these different principles is checked experimentally.

Note that it is possible to build the GS schemes based on complete graphs with any number of nodes. The external environment defines, ultimately, the course of evolution, not, however, through a simple link between hereditary variability of

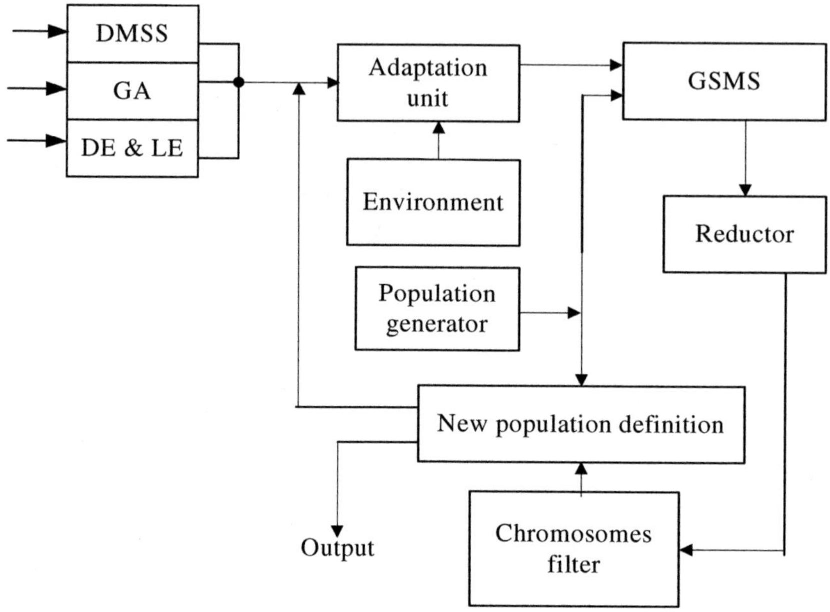

Fig. 2.12. The combined scheme of solution making based on GSMS

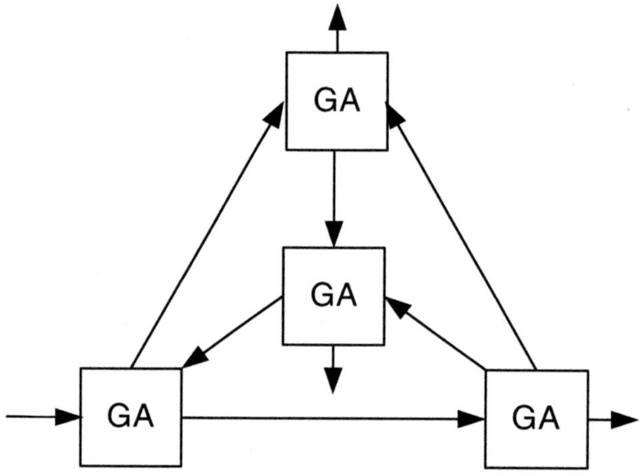

Fig. 2.13. (a) The search scheme based on tetrahedron

organisms and the environment as in LE, but in a more difficult form. The evolution, based on natural selection, on each of its steps creates new conditions of the genotypic environment. The natural selection fixes only those genotypes, which produce the phenotype appropriately adjusted to the given external environment.

Now, the main purpose of improving the quality of OT solutions through application of GA is be able to get out of the local optima, by the appropriate choice of

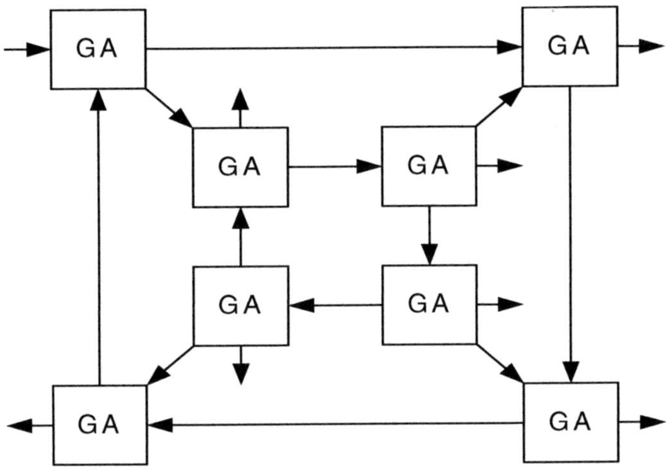

Fig. 2.13. (b) The search scheme based on hexahedron (cube)

GO, methods of selection, depth of the hierarchy of GSMS, the number of search generations, the size of population, etc. In OT of DM the non-standard architectures of GS are applied, with meta genetic optimization, as well as with migration and artificial selection. For construction of the basic framework of GA, we propose to use both synergetic and homeostatic principles of manipulating and adapting the AIS to the external environment. In some search schemes, it is possible to use models of joint evolution for finding acceptable quasi-optimal solutions.

2.2.2 Future-Oriented Technologies of Genetic Search for Solving the Optimization Problems

The artificial intelligence systems for optimization tasks of solution making are considered to be highly complex systems and one can hardly use a single model, scheme, or architecture representing a complex network of interconnected units, properties and functions, functioning in a certain interaction with an environment as such a system. Such models do not allow for carrying out of evolution by analogy without the use of a synthetic evolution theory (SET) and consideration of inconsistencies, with identification of the elements and relations that provide new knowledge on the object studied.

In order to enhance the quality of optimization tasks, in addition to the here described technologies, the ones connected with the multi sequencing GA, population structuring, chromosome migration, etc. are used [50]. For example, two approaches of population structuring are usually applied:

Chromosomes are divided into some subpopulations, randomly or in a predefined way, after the design of a population. DE, LE, FE or their modification is applied inside each subpopulation.

Chromosome location in a population is defined after the initialization of a population for every chromosome. Further, population is partitioned, if necessary.

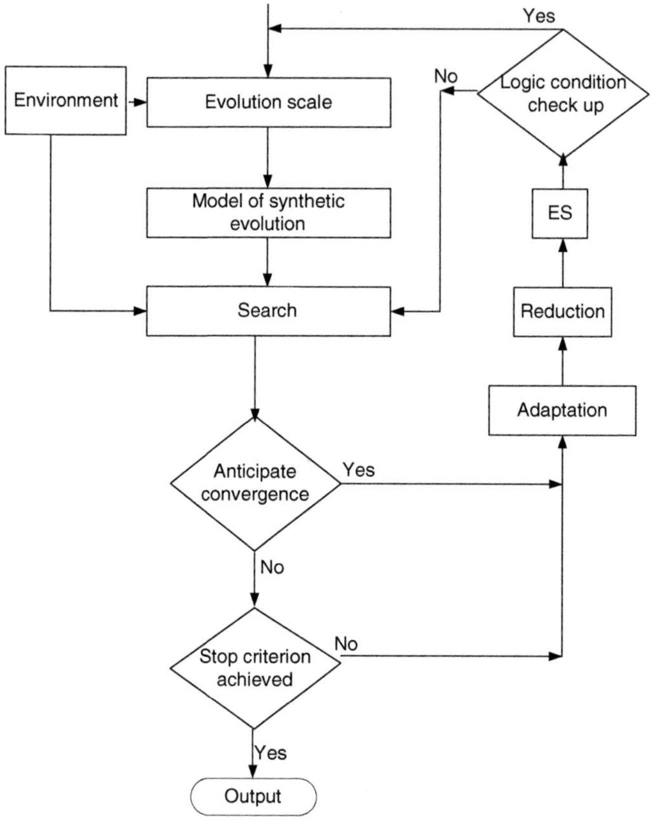

Fig. 2.14. Modification of the basic framework

The choices related to the use of GO depend on the "closeness" of solutions in space (based on TF value).

In Fig. 2.14 a modification of the basic GS framework is given, based on the use of SET. Here, in contrast to the basic framework, the scale of evolution, cooperating only with the external environment, produces signals as to the choice of the evolution model. One kind of heuristic search is realized through evolution. The blocks of adapting and ES are added in a chain of feedback. They allow for developing structures from chaos, establishment of balance in the system, choice of parameters for GA operation with the purpose of obtaining optimal and quasi-optimal solutions of the OT.

K. De Yong has also proposed five variants of planning the GA process [29, 69]:

- the elite model; let P'(t) be the best chromosome, generated at time t; if P'(t) has not remained in the pool after generation (t+1), P'(t) is introduced into the population at (t + 1) as N + 1;

- the model with the expected value; here, a chromosome is selected and the expected number of descendants for each population is calculated;
- the elite model with the expected value; this is an association of the first and the second model;
- the crowding factor (F); when the descendant chromosome has appeared after the generation of GA, it is checked for survival; the chromosomes with TF below the average are deleted. Usually, F = 2, 3, 4;
- the model of generation of a cross-over; in this case a number of CO points is produced.

The authors propose new variants of GA planning, based on the trial-and-error approach (synergetic-homeostatic-evolutionary), and combined evolution, adaptation of DM to the external environment, use of static optimization methods (SOM) and various retrieval strategies.

One of the simplest GS technologies, as opposed to SGA, is the so-called "steady reproduction" (steady-state reproduction) [34]. It works according to the following mechanism:

1. Create k descendants (chromosomes) from an initial population through reproduction.
2. Remove k members from the initial population to release the place for the descendants produced.
3. Evaluate and insert the obtained chromosomes (descendants) into the released places in the population.

A class of GA with directed mutation is proposed in [70]. They consist of two levels. On a node level the search is carried out for a group, and on the lower level the individual search is performed. On the first stage a population several times bigger than in SGA is used with the purpose of having a broader scope of search in the search space. Members of the population are evaluated, the chromosomes with TF less than average are discarded, and subpopulations are formed from the left chromosomes. Further search is carried out inside separate groups. Contrary to [70] it is proposed to group GA with a directed set of GO (CO, OSM, IO, SO, TO, and their various modifications) and GSMS. Let us outline the framework of the new algorithm:

1. Design an initial population P of the size |P| = N, Pi ∈ P, I = $\overline{1,N}$.
2. Determine TF for all chromosomes in a population and calculate the average TF as $f_{avg} = \frac{1}{N} \sum_{i=1}^{N} f(P_i)$.
3. At the stage of selection leave in the population the chromosomes with TF above the average for the population, P_j (j = 1, N), $f(P_j) > f_{avg}$.
4. Select DE, LE or FE with the help of a scale of evolution, based on the interaction with the external environment, blocks of adaptation, ES and GSMS.

5. Derive the groups of chromosomes by their spatial closeness, $|P_i — P_j| \leq \delta$, where δ is the neighborhood parameter: $\delta = 2L_k/N$, where L_k is the length of a segment, over which the k-component of P_i and P_j is defined. Thus chromosomes with the "useless" TF are deleted, and each group might investigate a separate local optimum.
6. Carry out a new selection, having left just one chromosome in each group forming the population partition.
7. Carry out GO (CO, OSM, IO, SO, TO) for all chromosomes as follows:

$$P_i^{t+1} = \begin{cases} P_i^t + w, if, f(P_i^t + w) > f(P_i^t) \\ P_i^t - w, if, f(P_i^t - w) < f(P_i^t). \\ P_i^t \text{ - otherwise} \end{cases} \quad (2.26)$$

Here, w is a random number from the interval $\left[o, \delta\left(1 - \frac{t}{T}\right)\right]$, where t is the number of GA generation, $t \in \overline{1,T}$; T is the predefined number of generations; δ is the neighborhood parameter, which does not permit chromosomes to quit a given group after realization of GO. GO will be carried out until the criterion of GA stop is satisfied. Formula (2.26) says that the more generations of GA, the less changes in chromosomes, i.e., the GS steps decrease. In another approach to creation of an initial population on a segment corresponding to chromosome length, the non-uniform grid with random step Δ, satisfying condition $\frac{0,5\,L}{N} \leq \Delta \leq \frac{2\,L}{N}$ is established. The degree of success of the proposed GA depends on the correct choice of the population size N. In many OT the number N is not known beforehand and it can only be forecasted, using knowledge on the way of functioning of the DM process, or any other relevant knowledge. In Fig. 2.15 the integrated scheme of the outlined GA is shown for the division of a population into two parts.

Here, in the blocks of the GO, the genetic operators CO, SMO, IO, SO, TO are realized in view of the expression 2.26. Deleting of chromosomes with TF below the average is carried out in the block of reduction. Selection of parameters of genetic search is made in the interaction with the external environment in EAB. Here, SET is, as before, a synthetic theory of evolution.

Let us consider the new technologies of construction of GO in the common framework of ES for solving OT. The function of recombination (RF) with multiple parents has been rarely used for solving technical problems until recently [34]. There are three various mechanisms of RF with multiple parents, called RF of the majority, RF of pairing, and RF of pairing on the average. All these operators can be used with any number of parents.

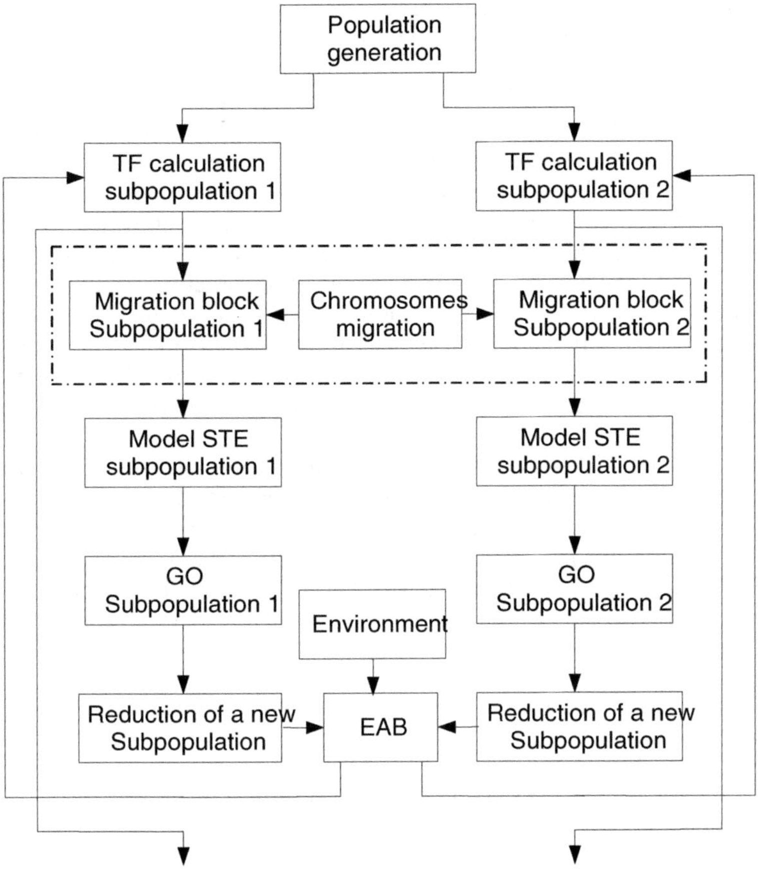

Fig. 2.15. The integrated scheme of the proposed GA

The global recombination is known in the evolutionary strategies since the 1970s [31—33]. More than two recombination operations are used in it, but just two chromosomes are, randomly, involved in production of each gene of the descendants. This allows for involving all the population in the RF. The multi chromosome mechanism of global recombination is a consequence of exhaustive search of chromosomes, therefore, more than two parents can take part in generation of descendants, and their numbers are not a priori defined. Let us note that a noticeable acceleration of search is observed for the transition from the asexual to sexual reproduction scheme (to the algorithm using only mutation, the algorithm of recombination of two parents is added), but only a small further increase is observed at the transition from the recombination of two parents to the recombination of several chromosomes (when global recombination is used instead of the variant with two parents).

It is possible to design an RF with a customized operation scale in the evolutionary strategies based on adaptation blocks and GSMS. The parameters (N, L, r) of the evolutionary strategies provide for a free choice of the number of parents. The parameter r defines the number of chromosomes and RF is applied to any set of r parents. The discrete version selects randomly one parent per chromosome. The average version gives the averages of all the parent chromosomes as chromosomes of the descendant. The number r is an independent parameter of recombination. It is necessary that all parents be different, randomly and uniformly selected from the population of dimension N for respective analysis. This mechanism selects and multiplies the blocks of appropriate genes of potentially various lengths. The new operator is applied after r parents are selected from the population of N individuals. The choice of two chromosomes for every case *i* is made only from a set of r individuals (parents). In this case the original mechanism is in a way preserved, as far as possible, though the variations between two extremes, r = 2 and r = N are possible.

The results considered above allow for the conclusion that the use of GO with several parents leads to certain increase in the speed of GA. However, this increase does not always take place. The quality of the concrete GA can be estimated through the distance $D = (f_1\text{-}f_2)/f_1$, where f_1 is the global maximum, and f_2 is the best value obtained by the algorithm. In other words, GA with operators of a higher level operate with larger populations, which explains their advantages. On this basis, we shall forward the following working hypotheses:

Hypothesis 1. Use of a smaller number of points of discontinuity in GO results in the increase of GA speed.

Hypothesis 2. Larger population sizes increase the quality of the GA.

Hypothesis 3. Use of a bigger number of parents yields better quality of GA.

The latter hypothesis is represented through two sub-hypotheses:

- the increase of the number of parents from one to two results in the increase of quality of the evolutionary algorithm;
- the increase of the number of parents beyond two results in the increase of quality of the evolutionary algorithm, but increases also the time of functioning of the GS.

Let us note that sub-hypothesis 1 represents the proposition that RF and SMO together work better than SMO alone. This allows for concluding that it is desirable to use RF with several parents in GA to solve the OT of DM though there is no guarantee of success.

As noted above, it is attempted to establish the synergetic rules, through which order is introduced into the chaotic systems, realized with AIS [71].

Benoit Mandelbrot developed fractal geometry of NS in the 1980s [71, 72]. It is proposed in [71, 72] that fractals fulfill the role of giving information and signal on the state of a system. The informational properties of fractals are close to those of a living cell. They are invariant to the object considered, are capable of self-similar reproduction in various spatial and temporal levels and transmitting information on the system about violation of stability of the structural state; have properties of adapting to external effect by self-modification of the fractal set (FS) at the points of instability of the system frame. For AIS, the self-similarity of frames, as a property of FS, is realized only on a limited scale.

For simulation of the self-organizing AIS, such as FS, it is promising to use the principle of subordination, when upon the transition of AIS through a point of bifurcation the set of variables is submitted to one or several variables, called the parameter(s) of the order. According to [73, 74], an evolving system has a fractal nature and develops not in a directed manner, but through spontaneous selection and self-consistent evolution.

Fractal objects are self-similar, i.e. their character does not undergo essential changes with change of their scale. The sets having such a property are treated as having geometrical (scale) universality. The transformations, which lead to such schemes, are the feedback processes with a large number of iterations, when the same operation is carried out repetitively, which makes it similar to SET and the ideas of Wiener [22, 75].

The sets of this kind are called fractal sets. The Cantor's set (CS) and the Sierpinski's carpet belong among them [72]. They have the geometrical invariance and are referred to as "the sets of the middle thirds". Thus, a segment of unit length [0, 1] is divided into three equal parts and the middle one, the interval [1/3, 2/3] is cut. The procedure is repeated recursively. We obtain a sequence of segments of decreasing length. In the first stage it is one segment, in the second - two, in the third – four, etc., in K-th - 2k. For $k \rightarrow \infty$ we have a point set referred to as MK. The total length of all cut segments is equal 1.

A new CO, based on CS is proposed. Let us call it a crossover operator based on Cantor set (COCS), and we shall show its operation on an example. Assume that there are two chromosomes P_1, P_2. According to the construction of CS, we shall cut in P_1, P_2 the segments [1/3, 2/3] and swap appropriate genes:

P_1 : 1 2 3 | 4 5 6 | 7 8 9

P_2 : 2 4 6 | 8 1 3 | 5 7 9

P'_1: – – – 8 1 3 – – –

P'_2: – – – 4 5 6 – – –.

Then, P'_1 is filled with the genes from P_1, from left to right, excepting the replicating and cut genes. The empty positions are filled by genes from P_2. A similar procedure is carried out for P'_2. Then we obtain:

P'_1: 2 7 9 8 1 3 4 6 5

P'_2: 2 5 7 4 5 6 9 1 3.

Continuing the iterative process, we cut in the chromosomes P'_1, P'_2 two segments between [1/9, 2/9] and [7/9, 8/9], and we swap appropriate genes. Thus, we obtain:

P'_1: 2 |7| 9 8 1 3 4 |6| 5

P'_2: 2 |5| 7 4 5 6 9 |1| 3

P''_1: – 5 – 8 1 3 – 1 –

P''_2: – 7 – 4 5 6 – 6 –

P''_1: 2 5 9 8 1 3 4 1 6

P''_2: 2 7 9 4 5 6 3 6 1

P''_1: 2 5 9 8 1 3 4 7 6

P''_2: 2 7 9 4 5 6 3 8 1.

So, using COCS we obtained four descendant chromosomes, P'_1, P'_2, P''_1, P''_2.

An example of a fractal object is the snowflake scheme. If we take an equilateral triangle and divide each of its sides into three parts and on each of three central thirds we construct the equilateral triangle of an appropriate smaller size, we obtain a fractal object. Note that it is possible to build any GO on the basis of CS. So, for example, MOMK consists in swapping the genes occurring behind the cut points. For example:

P_1 : 1 2 3 | 4 5 6 | 7 8 9

P'_1 : 1 2 7 | 4 5 6 | 3 8 9.

We can propose new algorithms for constructing GO, using the methods of one-dimensional search (passive, sequential, dichotomy, Fibonacci and gold section), based on the above principles.

The simplest of GO are CO, SMO, IO, SO, TO, and their modifications on the basis of the passive search, with random choice of chromosomes and of discontinuity points on them. For sequential search, the exhaustive search of the points of discontinuity for finding chromosomes with the optimal TF value is carried out. Development of GO using the method of dichotomy (DM) is done at the expense of the mechanism of exhaustive search of discontinuity points.

Let us formulate the integrated algorithm [76] of CO construction with the method of dichotomy (CODM).

1. Let two parent chromosomes of the length L be given.
2. Divide the chromosome (segment L) into halves (for an odd length – into near-halves).

3. The point of discontinuity defines the point of CODM.
4. Two new descendant chromosomes are obtained by the rules of OCO construction.
5. Each half of the descendant chromosome is bisected again and the process continues according to the initial scheme: $1BB1 \bigcup 2BB3 \bigcup 1BB2 \bigcup 2BB4$ for the first descendant chromosome; and for the second of the descendant chromosome: $2BB1 \bigcup 1BB3 \bigcup 2BB2 \bigcup 1BB4$. The process goes on until a given number of the descendant chromosome is obtained or DM is completed. Upon obtaining illegal chromosomes with the replicating genes, the respective correct genes are selected from one of the parental chromosomes to replace the last inserted illegal ones.
6. The algorithm terminates its functioning.

Consider an example. Let two parent chromosomes P_1, P_2 be selected for CODM, the GSMS block has been set to obtain four descendant chromosomes:

P_1 : 1 2 3 4 | 5 6 7 8 $\qquad$ P_3 : 1 2 | 3 4 | 8 6 | 7 5

P_2 : 4 2 1 3 | 8 6 7 5, $\qquad$ P_4 : 4 2 | 1 3 | 5 6 | 7 8,

---

P_3 : 1 2 3 4 | 8 6 7 5 $\qquad$ P_5 : 1 2 5 6 3 4 7 8

P_4 : 4 2 1 3 | 5 6 7 8, $\qquad$ P_6 : 4 2 8 6 1 3 7 5.

Four descendant chromosomes are obtained: P_3, P_4, P_5, P_6. Here, on the second step of DM, for the chromosome P_3 we have 1BB1 = (1,2), 1BB2 = (3,4), 1BB3 = (8,6), 1BB4 = (7,5), and for the chromosome P_4: 2BB1 = (4,2), 2BB2 = (1,3), 2BB3 = (5,6), 2BB4 = (7,8). The efficiency of the method of dichotomy grows exponentially with the number of chromosomes. We shall call the number of partitions in CODM the depth of dichotomy. It is usually defined by DMP, or is defined in AIS by the adaptation or GSMS blocks. Let us note that there are plenty of ways of concrete implementation of CODM. The expediency of CODM is defined on the basis of probability of survival, determined with the formulae (2.13), (2.14), or their modifications. Construction of other operators by the method of dichotomy is done in a similar manner.

Construction of the GO using the Fibonacci method (FM) is realized with a similar DM of the mechanism of exhaustive search of the points of discontinuity. Let us formulate the integrated schematic algorithm of construction of IO with the use of the Fibonacci method (IOFM). The property of the Fibonacci numbers (FN) is used in algorithm, namely that the next element in the FN sequence equals the sum of the two previous one, except for the first and second. Let a parent chromosome of length L be given.

1. The point of discontinuity of IOFM corresponds to the third FN.
2. We obtain the new descendant chromosome by inverting the part to the right of the IOFM point by the rules of construction of IO.

3. Further, 4th, 5th,... FN are selected as IOFM points and we pass to step 2. The algorithm ends operation upon the instruction form the DMP, based on indication from the adaptation and GSMS blocks or when the number FN ≥ L.

Let us consider an example:

P_1: 1 | 23456789

P_2: 1 | 98765432,

P_2: 198 | 765432

P_3: 198 | 234567,

P_3: 19823 | 4567

P_4: 19823 | 7654.

Here, P_1 is the parent chromosome, and P_2, P_3, P_4 are the descendant chromosomes. The expediency of IO of FM is defined on the basis of probability of survival, determined by the formula (2.36), or by its modification. Other GO based on FM are developed similarly.

An important question in the implementation of GS is constituted by the construction of TF. Let us outline five main methods of construction of the TF. In the first method, the TF $f(P_i)$ of a chromosome $P_i \in P$ is defined in the natural terms of OT of DM, based on decimal coding. For example, the cost of partitioning of a graph is estimated in the decimal system. In the second method the TF $f(P_i)$ of $P_i \in P$ is defined on the basis on binary coding. The Gray and Hamming codes can be used here [34]. In the third case the standard TF: $f_c(P_i) = f(\max) - f(P_i)$ for a case of maximization and $f_c(P_i) = f(P_i) - f(\min)$ for minimization are used. Here $f(\max)$ is the maximum value of TF, and $f(\min)$ is the minimum value of TF. In the fourth method an updated TF value $f_m(P_i) = 1/(1 + f_c(P_i))$ is calculated. The value of this TF lies in the interval [0, 1]. In the fifth case we introduce the normalized TF $f_n(P_i) = f_M(P_i) \Big/ \sum_{i=1}^{N} f_M(P_i)$, where N is the size of the population. Further versions of TF can be defined on the basis of modifications and various combinations of the here given ones.

In the conclusion of this subsection let us summarize. Various techniques of solving OT of DM using SET were described. They can allow for expanding the region of search for solutions without increasing the operating time of algorithms and for reducing the chances of premature convergence of algorithms. The concepts of using fractals, as well as the methods of search involving dichotomous partitioning, Fibonacci numbers and gold cut, are described.

To conclude, let us state that there are three types of the mechanism of creation of multi-chromosome search operators:

- the GO based on the repetition of the parent chromosomes;
- the GO based on segmentation and recombination of the genetic information from the parents;
- the GO based on numerical analysis of TF.

It cannot in general be expected that three different algorithms will produce identical results at various operator levels. The advantage of all these methods is that they allow for reduction of the uncertainty interval i.e. the search area.

2.3 Optimization Problems of Decision-Making on Graphs

2.3.1 The Analysis of Genetic Algorithms of Graph Partitioning on the Basis of Artificial System Models

The formalization of the problem of AIS reduces to various combinatorial OT of the theory of graphs, like the task of graph partitioning, graph coloring, the Traveling Salesman, construction of the maximal dominant and independent sets, cliques, etc. The stages of the respective DM process are as follows:

- the preliminary analysis and identification of problems;
- the statement of the task to be solved;
- the acquisition of the initial data;
- the data processing and finding of solution to a concrete task;
- the analysis of the obtained solution.

The choice of typical AIS representation of data is connected, on the one hand, with the process of evaluating the alternatives by the DMP, and on the other hand - with data processing. AIS are usually represented by graphs. Transformation of graphs consists in ordering of nodes, identification of subsets of nodes (alternatives), definition of partitions, accommodation, coloring, matching, nuclei, clusters etc.

The dialogue with the DMP takes place at every stage of DM, at the choice of a suitable model, and solving of the combinatorial task of optimization, with the management and correcting of solution strategy being carried out. After formulation of a concrete OT of DM, the question of model choice is put, identified with a combinatorial task of discrete optimization. Then, it is possible to interpret DM as a process of selection of the best approximations from a given class. The basis for this selection is constituted by the principles of ordering proper for the AIS from the class considered. Thus, various models of solving the same OT DM will differ as to this kind of principles forming their basis.

Let us assume that a set of AIS is examined and certain OTDM is investigated, with the solution process understood as the search for optimum approximation of the initial AIS within some definite class. Then it is possible to say that the model of solving a given OTDM is defined on the set of initial AIS if a certain principle or rule is specified, according to which to any matrix or graph of AIS some subset of alternatives corresponds. The definition of a principle of a selection of optimum alternatives reduces to a sorting task formalized as a combinatorial OT on graphs [66].

One of major OTDM is to partition the graph into a given or any number of parts [75, 77-84]. The task of graph partitioning has many practical applications. It is used in designing devices of automatics and computer equipment, development of control systems, computer and engineering networks, and also in solving various AI tasks. Let us note that the task of graph partitioning belongs to the class of the NP-complete problems, i.e. there are no effective algorithms to solve it with polynomial time complexity.

The classification of the existing methods of graph partitioning is given in [75]. For solving the OT of large dimensions of greatest interest, from the point of view of implementation, are the sequential and iterative heuristic methods. Time complexity of such methods is of the order O (n) — O (n^3), where n is the number of nodes in the graph. The random search methods can be used in some cases for graph partitioning, but they do not guarantee obtaining of a local optimum in a finite number of steps. The methods of GS found recently wide application in solving the graph partitioning problems [24, 38, 84-87]. These methods use ES and the criterion of "the survival of the strongest" to obtain the optimum solutions. The methods described in the first section, for partitioning the graph into subgraphs, are the new and modified GA with management of the search process on the basis of synergetic and homeostatic principles. They differ from the previous ones by application of the non-conventional GS architecture, and the new GO exploiting knowledge of the task solved.

Let us formulate the statement of the problem of graph partitioning into a given or any number of parts. Let the graph G = (X, E), be given, where X represents the set of nodes and E is the set of edges.

Let B = $\{B_1, B_2 \dots B_s\}$ be the set of parts, resulting from partitioning of the graph G into the parts $B_1, B_2 \dots B_s$, such that $B_1 \bigcap B_2 \bigcap \dots \bigcap B_s = \varnothing$, and $B_1 \bigcup B_2 \bigcup \dots \bigcup B_s = B$. Let each part B_i consist of elements b_m, $B_i = \{b_1, b_2 \dots b_n\}$, $n = |X|$. Then the problem of partitioning a graph G consists in determining the parts $B_i \in B$, satisfying the following three conditions:

$$(\forall B_i \in B)\ (B_i \neq \varnothing),$$

$$(\forall B_i, B_j \in B)\ ([B_i \neq B_j \rightarrow X_i \bigcap X_j = \varnothing] \wedge$$

$$\wedge\ [\ (E_i \bigcap E_j = E_{ij}) \vee (E_i \bigcap E_j = \varnothing)]), \tag{2.27}$$

$$\bigcup_{i=1}^{s} B_i = B, \bigcup_{i=1}^{n} E_i = E, \bigcup_{i=1}^{n} X_i = X, |E_{ij}| = K_{ij}.$$

The target function for graph partitioning is defined as follows:

$$K = \frac{1}{2}\sum_{i=1}^{n}\sum_{j=1}^{n} K_{i,j}, \ (i \neq j) \tag{2.28}$$

where $K_{i,j}$ is the number of connections between parts B_i and B_j resulting from partitioning of the graph G; K is the total number of edges involved in partitioning of the graph.

The standard problem of partitioning consists in minimization of K (K→min). Minimization of K in graph partitioning allows, indirectly, for taking into account of many criteria and construction-technological constraints in the AIS model for solving the OTDM. Hence, the problem of partitioning consists in the search for such a partitioning into B_i in the set of possible partitions of the graph, which minimizes the value of K, being the TF of partitioning, with all the respective constraints accounted for.

Graph partitioning belongs to the class of problems of combinatorial optimization on graphs. The total number of solution variants is equal to the number of rearrangements of n graph nodes, i.e. $C_n = n!$, and in view of constraints on formation of subsets (parts of the graph) - the number of combinations of n nodes into m subsets [75], i.e.:

$$C_n^m = n!/m!(n-m)!. \tag{2.29}$$

It follows from the above that the solution to the graph partitioning problem on the basis of exhaustive search is inconvenient because of the exponential complexity of such a process. In this connection, various heuristics have been developed for solving this problem [41]. The iterative methods of pair and group rearrangements, methods of consecutive approximation, and relaxation, of depth-first and width search, directed exhaustive search etc. belong among them. Recently, various random methods have been appeared among the heuristics. These methods are ES, MO, GS and their modifications. All of them are based in a sense on the analysis and exhaustive search of solution variants. They differ, though, in technology and principles of realization.

The methods proposed do not guarantee finding of the optimum solution and strongly depend on the class and description of the problem solved. The heuristic methods allow in practice for receiving quite good results, but it is impossible to predict their behavior. They can be used independently, and together with other methods. This allows for reducing time of operation. The respective ART (algorithm running time) can be of any magnitude, but the majority of known algorithms have ART between O (n) and O (n^2).

The essence of the majority of algorithms of graph partitioning consists in the choice of some initial partitioning and its subsequent improvement with the help of an iterative, pair or group exchange of nodes between various parts in the partition. The rearrangement of nodes is carried out at each iteration, so as to maximally reduce the number of connections between the parts of graph or minimally increase the number of external edges.

The following algorithms are among the most known ones used in graph partitioning:

- those proposed by A. N. Melikhov, L. S. Bershtein, V. M. Kureichik; V. Kernigan, S. Linn and their improved heuristics [41, 82];
- formation of the minimal sets [75];
- simulated annealing [34];
- of S. Fiducceo, R. Mateuss and their modifications [83];
- matrix-based [84];
- of random assignments [79];
- of evolution simulation [24, 37, 38, 88].

The comparative evaluation of the existing methods shows that the algorithms using highly effective heuristics give approximately the same result, spending at each generation much less (up to 10 times) time. This is connected with the fact that many of the heuristics do not use the complete exhaustive search, but a directed improvement of TF, with rejection of the unpromising solutions. The GA based on the ES methods offer this kind of advantage.

Let us provide a brief description of the algorithms using various heuristic bases. Among the popular algorithms of graph partitioning are the iterative algorithms [41, 84], requiring some initial partitioning. They are relatively sensitive to initial partitioning, which influences the operating time of the algorithm. Besides, they frequently get stuck in a local optimum, when the size of the graph is large ($n > 1000$). One of the ways to overcome this shortcoming consists in grouping of the strongly connected nodes into clusters to then include these clusters in the parts of the graph, and in using fractal sets for grouping the strongly connected nodes [89].

A. Melikhov and associates, V. Kernigan and S. Linn proposed an iterative heuristic algorithm independently, this algorithm starting from a random partitioning and then applying exchange in pairs between two subsets, improving the initial solution by the repeated rearrangement of nodes.

The essential advantage of GA is that fact that they can work for any class of problems on graphs. In this context it is important to present problem formulation in the genetic form correctly so as to map the TF search space properly. For this purpose the problem analyzed is considered from the point of view of the criteria used and the optimal choice of ES, which, in turn, defines the coded representation of the solutions in GA.

Let us construct the scheme of graph partitioning on the basis of GS. Such a scheme is given in Fig. 2.16. Here, in the first stage the current (initial) population of solution is constructed, i.e. the subsets B_1, B_2, … B_k ($k \leq s$) are determined. Construction of the population is influenced by the external environment of DMP or ES. This set of solution candidates can be obtained by the random, directed or random-directed methods. In cases, when the number of elements is large, it is most preferable to use sequence, iterative, random and search algorithms for producing the population. In the second stage measurement of TF is done with K (expression (2.28)). There are studies [41, 84], in which it is proposed in solving the

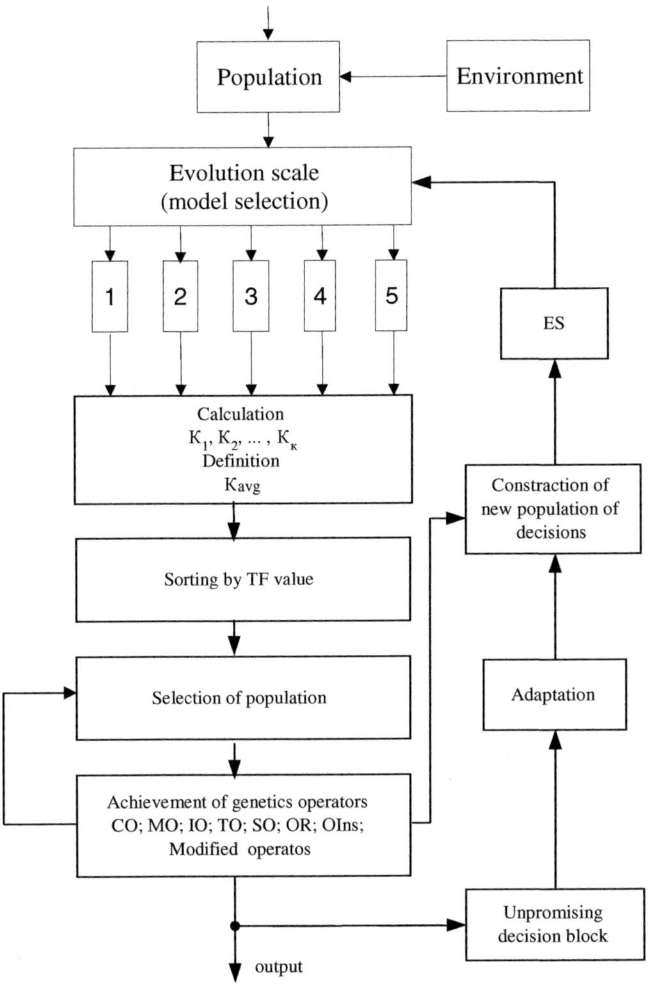

Fig. 2.16. A modified GS scheme for the partitioning problem

graph partitioning problem to not minimize K (number of connections between the parts forming partition), but to maximize the number of connections M inside these parts. Blocks 1-5 correspond to different genetic operators (1-CO, 2-MO, 3-SO, 4-IO, 5-TO). At stage 3 we calculate K for each element of the current population after genetic operators use. We can define K_{avg} , the average value of TF for the given population. At stage 4 in Fig. 2.16 sorting of the population on the basis of K is performed. All of the methods of selection, described above, can be applied for this. Elements featuring the least value of K are placed first, etc. Stage 4 performs selection of the population to obtain parental pairs. This selection is carried out by one of methods described earlier.

The elementary case of selection can be the consecutive choice of the best (in the sense of function K) chromosome after sorting. Stage 5 implements GO and their updating again. At stage 6 (Unpromising decision block) unpromising decisions are collected. Stage 7 implements the strategy of adaptation and on the basis on feedback selects the evolution algorithm or the respective modification, and orders the use and application of various GO. At the eighth stage construction of a new population of solutions is carried out. The ninth stage operates the process of search with the help of information feedback based on expert system (ES). Then, for every chromosome from the new population the value of K is calculated and those elements from the old and new population survive, for which $K \geq K_{avg.}$. The number of elements in a new population should not exceed the number of elements in the old population. It is possible to propose a lot of similar GS schemes for solving the partitioning problem.

Let us consider the consecutive GA of partitioning. Let graph G = (X, E) be given. At the beginning we order all nodes the graph according to the increasing values of local degrees of nodes. This corresponds to the list of nodes, and can also be treated as corresponding to a trivial partitioning, when the groups forming partition are equivalent to the ordered nodes of the graph. Then, we begin to form pairs of nodes and to calculate the index of coherence for each pair:

$$\delta_{i,j} = e_{i,j} - (e_{i,t} + e_{j,t}), \qquad (2.30)$$

where $e_{i,j}$ is the number of edges between nodes x_i and x_j, forming a pair; $e_{i,t}$ and $e_{j,t}$ are numbers of edges connecting the chosen pair to all other nodes of the graph, and t = 1, 2… n-2. Let us make the second list, containing the pairs of nodes, for which $\delta_{i,j} \geq 0$. Note that when there are no pairs with $\delta_{i,j} \geq 0$, two cases are possible. In the first case, the pairs with the least negative $\delta_{i,j}$ are formed. In the second case the pairs are not formed, and a transition to formation of triples of nodes takes place. Further on the process is repeated in a similar way, until the partitioning of the graph into the given or any number of parts is fully executed. This is the basic idea of the standard procedure of partitioning [41]. For a small number of nodes it gives satisfactory results, since it practically ensures the complete exhaustive search in solution space. The use of such an approach becomes inconvenient with the increase of the number of nodes. A simple and modified GO can be applied to reduce the number pairs considered. Let us note that it is possible to use the following expression:

$$\tau_i + \tau_j \geq \tau_t, \qquad (2.31)$$

where $\tau_i + \tau_j$ are the internal edges of the partition, and τ_t (t=1, 2… n-2) are the external edges of the partition.

Consider an example. Let a graph be given (Fig. 2.17). Define the local degrees of nodes and order them in the increasing sequence (Table 2.1). Then, construct pairs of nodes to define partition of the graph into parts with two nodes in a part.

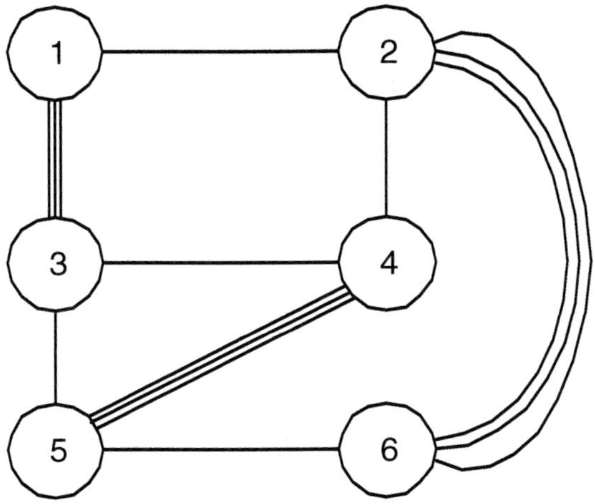

Fig. 2.17. Graph G

We start with defining two blocks in such partition. These are: (x_1, x_3), (x_2, x_6). (For the sake of simplicity we shall write sometimes the numbers of nodes (1, 2, …) instead of (x_1, x_2, …)). Now, if it were necessary to partition the graph into the pairs of nodes in its entirety, by minimizing K, it would be possible to take as the third block of nodes the one with the smallest negative value of the expression (2.31). Such partition is shown in Fig. 2.18.

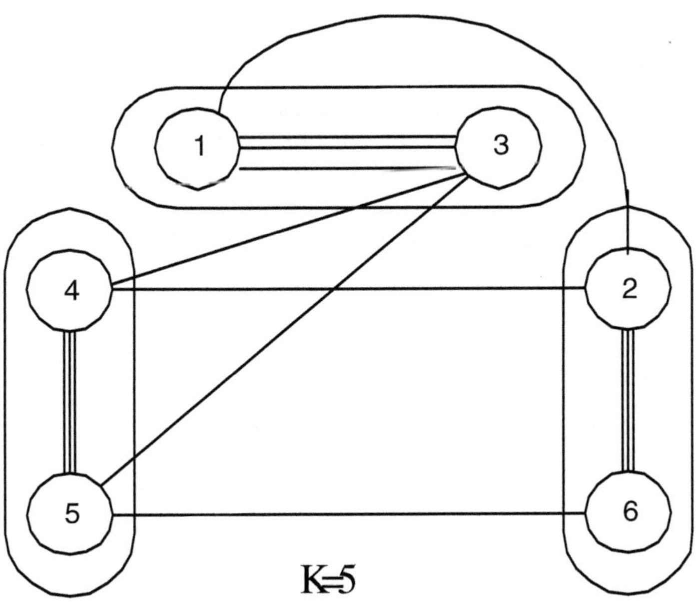

Fig. 2.18. Graph G partitioned into three parts

In Table 2.2 all pairs of nodes of the graph are characterized in terms of expression (2.31), and on this basis an appropriate selection can be performed. Obviously, for large graphs this procedure is inconvenient. Let us apply, therefore, one of the modified GS schemes, described previously.

Table 2.1.

X_i	X_2	X_3	X_4	X_5	X_6	X_1
ρ_{xi}	5	5	5	5	4	4

Table 2.2.

$(x_i\ x_j)$	(1,2)	(1,3)	(1,4)	(1,5)	(1,6)	(2,3)	(2,4)	(2,5)	(2,6)
$\tau_{i+}\tau_j \geq \tau_t$	$1 \geq 7$	$3 \geq 3$	$0 \geq 9$	$0 \geq 9$	$0 \geq 8$	$0 \geq 10$	$1 \geq 8$	$0 \geq 10$	$3 \geq 3$
$(x_i\ x_j)$	(3,4)	(3,5)	(3,6)	(4,5)	(4,6)	(5,6)			
$\tau_{i+}\tau_j \geq \tau_t$	$1 \geq 8$	$1 \geq 8$	$0 \geq 9$	$3 \geq 4$	$0 \geq 9$	$1 \geq 7$			

We start with constructing randomly a population of solutions, P = {P_1, P_2, P_3, P_4, P_5, P_6}, with P_1: (1, 2). We shall now check condition (2.31) for all the elements of the population P. For the first element: 1≥7 (condition does not hold). For the remaining elements we obtain:

P_1: (1, 2) — $1 \geq 7$, P_4: (4, 5) — $3 \geq 4$, P_5: (5, 6) — $1 \geq 7$,

P_2: (2, 3) — $0 \geq 10$, P_3: (3, 4) — $1 \geq 8$, P_6: (3, 6) — $0 \geq 9$.

Here, expression (2.31) was taken to express TF. Let us determine the value of TF for all chromosomes in the population. By ordering this population we obtain:

P_4, P_5, P_1, P_3, P_6, P_2.

Let us make selection and form parental pairs:

P_4: (4, 5) P_1: (1, 2) P_6: (3, 6)

P_5: (5, 6), P_3: (3, 4), P_2: (2, 3).

Let us apply standard CO and obtain new chromosomes:

P_4: (4, 5) P_1: (1, 2) P_6: (3, 6)

P_5: (5, 6) P_3: (3, 4) P_2: (2, 3)

———, ———, ———

P'_4: (4, 6) — $0 \geq 9$ P'_1: (1, 4) — $0 \geq 9$ P'_6: (3, 3)

P'_5: (5, 5) P'_3: (3, 2) — $0 \geq 10$ P'_2: (2, 6) — $3 \geq 3$ (Yes).

Due to a single application of CO we obtained the best solution P'_2 (2, 6). This solution corresponds to the first subgraph with nodes 2 and 6 (fig.2.18). The advantages from application of GA come out clearly even on a simple example.

Use of standard CO can yield infeasible solutions (partitions P'_5, P'_6). In this case, it is possible to propose a modification of the CO. This modification would involve analyzing solutions produced through the CO. The repeated elements would be eliminated, and the absent ones added. Then, in our example we shall obtain:

P''_4: (4, 5), P''_1: (1, 3), P''_6: (3, 2).

P''_5: (5, 6), P''_3: (2, 4), P''_2: (6, 3).

Here, the pairs with repeated parents are (P''_4, P''_5), and (P''_6, P''_2), and there is an element P''_1 with one of the best values of the target function ($3 \geq 3$). Element P''_1 corresponds to the second subgraph containing nodes 1 and 3. Continuing similarly, we shall construct groups for partitioning with three nodes each.

In Fig. 2.19 partition of graph G from Fig. 2.17 is shown into two parts with three nodes in each part. Here, TF = 5. Note that the procedure of genetic search can be applied both as unique and multiple optimization. In the first case, the complexity of the algorithm of graph partitioning is of the order of O (n) — O (n^3). The extreme case of O (n^3) will take place for the population of more than one hundred and for more than a thousand generations of the algorithm. Besides the GO and the search operators, we can insert into the GS scheme auxiliary blocks to increase the quality of the resulting algorithms.

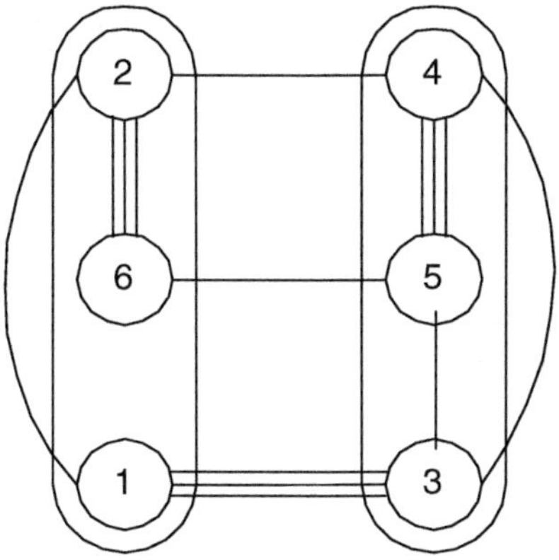

Fig. 2.19. Partition of G into two parts

As noted before, we usually consider fractals as sets with scale invariance, i.e. on any scale they look practically the same. The solutions to the majority of nonlinear OTDM represent fractals. In particular, strange attractors can have geometrical invariance, i.e. their structure, like in fractal sets (FS), repeats itself with the successive scale increases [73, 74].

Let us consider the aggregation mechanism (AM), describing the creation of fractals. According to AM, certain versions of fractals can be obtained during the disorderly, ordered or combined growth. For example, a cluster (an object featuring maximum internal coherence, meaning, in a graph - a subset of nodes or a clique connected internally as much as possible) can grow as follows: with time, some elements join it as if "sticking" to it. This process is referred to as aggregation. We can assume that particles join the growing cluster randomly.

Let us consider an algorithm of graph partitioning, based on a modified aggregation of fractals. The mechanism of graph partitioning is based on four principles:

- construction of clusters (sets of graph nodes connected among themselves);
- factorization of clusters, i.e. reduction of the dimensions of the graph by representing clusters as nodes of the reduced graph);
- AM, i.e. aggregation, proceeding by random, directed and combined association of nodes of the graph into clusters;
- use of the models of evolution.

The process of cluster creation is based on the concept of construction of minimal and quasi-minimal sets in the graph or hypergraph. A cluster is a part of a graph $\Phi \subseteq G$, $\Phi = (X_1, U_1), X_1 \subseteq X, U_1 \subseteq U$. The nodes of a cluster are connected by the external edges to other nodes $X \backslash X_1$ of the graph G. These external edges do not belong to the subset of cluster edges. We shall denote by f the cardinality of the subset of external cluster edges. A cluster Φ_i is referred to as minimal if for any other cluster Φ_j, $\Phi_j \subset \Phi_i$ the condition $f_i \leq f_j$ is satisfied. By definition we shall count:

$$(\forall \Phi_i \subset G)\ (\Phi_i \neq \varnothing),\ (\forall_{xi} \in X)\ ({}_{xi}\text{—minimal cluster}). \tag{2.32}$$

This means that a minimal cluster (CM) cannot be empty. Besides, the special trivial case is that each node a graph forms CM. We shall call a cluster quasi-minimal, if the following condition is satisfied:

$$f_{i+}\varepsilon \leq f_j, \tag{2.33}$$

where ε—is the factor determining to what extent it is possible to increase the number of external edges in the minimal cluster $\Phi_{i.}$ It is defined in the analysis of a mathematical model (MM) of the graph, on the basis of the decision of DMP or ES.

Theorem 2.1. If Φ_i and Φ_j are both CM and $\Phi_i \not\subset \Phi_j \wedge \Phi_j \not\subset \Phi_i$, then $\Phi_i \bigcap \Phi_j = \varnothing$.

Theorem 2.2. Let Φ_i, Φ_j be CM. If $\Phi_s = \Phi_i \bigcup \Phi_j \wedge f_s < f_i \wedge f_s < f_j$ then Φ_s is a CM. If $\Phi_s = \Phi_i \bigcup \Phi_j \wedge f_s + \varepsilon \le f_i \wedge f_s \le f_j$ then Φ_s is a QCM.

Theorem 2.3. Let Φ_1, $\Phi_2 \ldots \Phi_L$ be CM, and any association of a subset of them not be CM. If $\Phi_i = \Phi_1 \bigcup \Phi_2 \bigcup \ldots \bigcup \Phi_L \wedge f_i < f_1 \wedge f_i < f_2 \ldots \wedge f_i < f_L$ then Φ_i is a CM. If $\Phi_i = \Phi_1 \bigcup \Phi_2 \bigcup \ldots \bigcup \Phi_L \wedge f_i + \varepsilon \le f_1 \wedge f_i + \varepsilon \le f_2 \ldots \wedge \wedge f_i + \varepsilon \le f_L$ then Φ_i is a CM.

We shall describe a modified heuristic algorithm of construction of CM and QCM based on these theorems.

1. Order all the local degrees of nodes the graph as trivial CM and QCM.
2. Analyze in pairs all nodes of the graph G starting from nodes with the highest local degree. For a pair of nodes (x_i, x_j) such that:

$$f = \rho(x_i) + \rho(x_j) - 2r_{i,j}, \tag{2.34}$$

we have that $\Phi = (X_1, U_1)$, $X_1 = \{x_i, x_j\}$ is CM. Here $\rho(x_i)$, $\rho(x_j)$ are the local degrees of nodes x_i, x_j; $r_{i,j}$ - number of edges connecting nodes x_i, x_j among themselves, f is the number of edges connecting CM with other nodes of the graph G, and $f < \rho(x_i)$, $f < \rho(x_j)$. For a pair of nodes (x_i, x_j) such that:

$$f = \rho(x_i) + \rho(x_j) - 2r_{i,j} + \varepsilon, \tag{2.35}$$

and $f + \varepsilon \le \rho(x_i)$, $f + \varepsilon \le \rho(x_j)$, we have that $\Phi = (X_1, U_1)$, $X_1 = \{x_i, x_j\}$, is QCM. The CM or QCM Φ is entered on a special list, and nodes x_i, x_j are excluded from consideration. Step 2 is repeated, unless no new CM or QCM was formed, then pass to step 3.

3. Perform factorization. The nodes x_i, x_j in CM or QCM are replaced by one $x_{i,j.}$ Thus the edges connecting x_i and x_j disappear, and the edges from the nodes x_i, x_j to other nodes are attributed to the new node $x_{i,j.}$ A new graph G' is produced.
4. In graph G' all nodes are analyzed. If for a triple of nodes x_a, x_b, x_c there is:

$$f = \rho(x_a) + \rho(x_b) + \rho(x_c) - 2r_{a,b} - 2r_{b,c} - 2r_{a,c}, \tag{2.36}$$

and $f < \rho(x_a)$, $f < \rho(x_b)$, $f < \rho(x_c)$, the nodes x_a, x_b, x_c form a CM, according to Theorem 3.3. If for a triple of nodes x_a, x_b, x_c there is:

$$f = \rho(x_a) + \rho(x_b) + \rho(x_c) - 2r_{a,b} - 2r_{b,c} - 2r_{a,c} + \varepsilon, \tag{2.37}$$

and $f + \varepsilon \leq \rho(x_a)$, $f + \varepsilon \leq \rho(x_b)$, $f + \varepsilon \leq \rho(x_c)$, the nodes x_a, x_b, x_c form a QCM, according to Theorem 3.3. Pass to step 2. The CM or QCM are entered on the list and pass to 5. If a CM or QCM is not established, repeat step 4.

5. Increase the parameter of cardinality of CM or QCM by one and pass to step 3. If the cardinality of CM or QCM is equal to that of the graph analyzed, pass to step 6.
6. Termination of the algorithm.

Here, ε is the smallest relative deviation from CM ($\varepsilon = 0, 1, 2 \ldots$).

If, after the set of CM or QCM has been constructed, there are still free nodes in the graph G, it is necessary to carry out the aggregation procedure. It consists in the association of nodes in CM or QCM with an analysis of TF values.

Consider the example shown in Fig. 2.20.

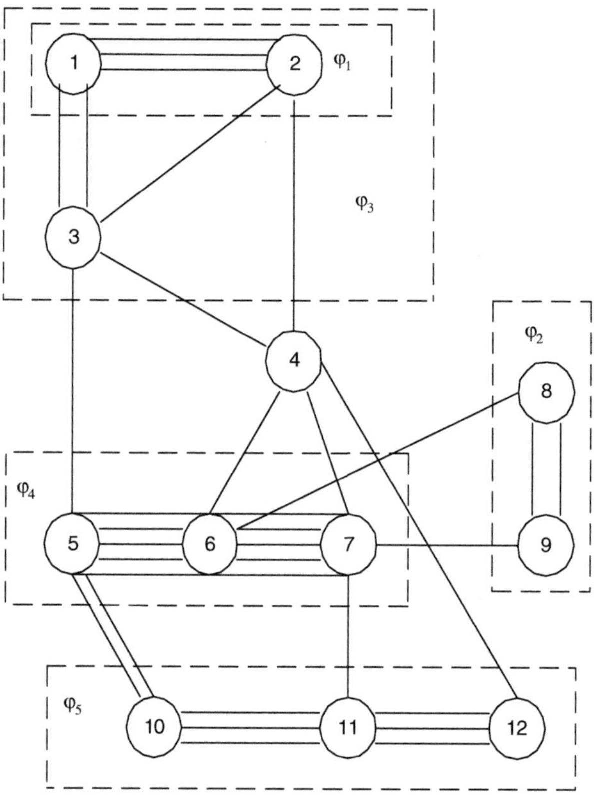

Fig. 2.20. A graph G

The matrix of contiguity of the graph G is as given below:

R =

	1	2	3	4	5	6	7	8	9	10	11	12
1	0	3	2	0	0	0	0	0	0	0	0	0
2	3	0	1	1	0	0	0	0	0	0	0	0
3	2	1	0	1	1	0	0	0	0	0	0	0
4	0	1	1	0	0	1	1	0	0	0	0	1
5	0	0	1	0	0	5	0	0	0	2	0	0
6	0	0	0	1	5	0	5	1	0	0	0	0
7	0	0	0	1	0	5	0	0	1	0	1	0
8	0	0	0	0	0	1	0	0	2	0	0	0
9	0	0	0	0	0	0	1	2	0	0	0	0
10	0	0	0	0	2	0	0	0	0	0	3	0
11	0	0	0	0	0	0	1	0	0	3	0	3
12	0	0	0	1	0	0	0	0	0	0	3	0

We will determine the local degrees of nodes from the matrix and we will construct the list of degrees of these nodes in the increasing order. Resulting from the performance of the algorithm, we shall designate as CM $\Phi_1 = \{x_1, x_2\}$ and $\Phi_2 = \{x_8, x_9\}$. Upon having treated Φ_1 and Φ_2 as nodes, we obtain a new graph, whose matrix of contiguity is as follows

R_1 =

	(1,2)	3	4	5	6	7	(8,9)	10	11	12
(1,2)	0	3	1	0	0	0	0	0	0	0
3	3	0	1	1	0	0	0	0	0	0
4	1	1	0	0	1	1	0	0	0	1
5	0	1	0	0	5	0	0	2	0	0
6	0	0	1	5	0	4	1	0	0	0
7	0	0	1	0	4	0	1	0	1	0
(8,9)	0	0	0	0	1	1	0	0	0	0
10	0	0	0	2	0	0	0	0	3	0
11	0	0	0	0	0	1	0	3	0	3
12	0	0	0	1	0	0	0	0	3	0

.

As there are no more CM of cardinality 2, we shall construct, according to the algorithm, the CM of cardinality 3. We obtain $\Phi_3 = \{\Phi_1, x_3\}$, $\Phi_4 = \{x_5, x_6, x_7\}$,

$\Phi_5 = \{x_{10}, x_{11}, x_{12}\}$. The algorithm terminates its functioning, there are no more CM. It is possible to add node x4, which is a trivial CM, to all other CM, selecting the proper ones through calculation of TF. We obtain, as a result, two variants of partitioning:

1st variant: $\{x_1, x_2, x_3\, x_4\}, \{x_5, x_6, x_7\}, \{x_8, x_9\}, \{x_{10}, x_{11}, x_{12}\}$, with $K = 9$.

2nd variant: $\{x_1, x_2, x_3\}, \{x_4, x_5, x_6, x_7\}, \{x_8, x_9\}, \{x_{10}, x_{11}, x_{12}\}$, with $K = 9$.

For example, if partition of the graph into three parts is given, variant 1 after the procedure of evolution will get the following form:

3-rd variant: $\{x_1, x_2, x_3\, x_4\}, \{x_5, x_6, x_7, x_8, x_9\}, \{x_{10}, x_{11}, x_{12}\}$, with $K = 7$.

Let us define the asymptotic estimate of the computational burden of this algorithm. Let N_p be the size of the population, N_d - the number of descendants generated, N_G - the number of GA generations. Then the general computational burden of the algorithm can roughly be defined as:

$$T \approx [\alpha N_p t_p + \beta N_g t_g + \gamma (N_p + N_g) t_c] N_G, \tag{2.38}$$

where t_p is effort required to construct one chromosome in a population; t_g is the effort needed to produce one generation of descendants; t_c — is the cumulative computational burden of selection; α, β, γ — are the factors determining the parameters of GS. The running time of the described algorithm is roughly of the order of $O(n^2)$ for one generation of GS.

Let us consider the basic idea of the iterative partitioning of a hypergraph $H = (X, E)$ with minimization of K. It is possible to define the cutting function for each edge $e_i \in E$, as number of points belonging to various blocks. If even one node of the hypergraph edge does not belong in the part with the other nodes of this edge, at least one cutting edge will necessarily exist. The sum of all the cutting edges shall define the overall number K of connections between blocks, which is being minimized.

Graph partitioning with application of GA always allows for obtaining a local optimum, for having a chance to leave it, and for getting nearer to the optimum and quasi-optimum solutions. In addition, it is especially important that the running time of the algorithm does not leave the area of polynomial complexity (ART $\approx O(n \cdot \log n)$ — $O(n^3)$).

2.3.2 The Genetic Approach to the Placement of the Graph Nodes

Formulation and solving of the OT on graphs allows for simulating various branches of AIS evolution. One of such OT is the placement of the graph nodes on a plane, on a line, within a volume and area of any arbitrary configuration on the basis of certain criteria of optimization.

According to [84, 90—95] it is possible to divide the algorithms of placement of the graph nodes into two basic classes. These are the constructive algorithms and the methods of iterative improvement. Six basic approaches are being applied in solving these problems [41]. They include MO; the force-driven placement; placement by means of partitioning; numerical optimization approaches; GA-based placement; and exact methods.

Let us formulate the problem here considered. The basic purpose of the placement algorithms is to minimize the total length of edges of the graph or hypergraph. Frequently the problem is formulated as (or restriction also is put) the one of minimizing the total number of crossings of graph edges (or the corresponding constraint is formulated).

A model for the problem of placement on a plane is shown in Fig. 2.21.

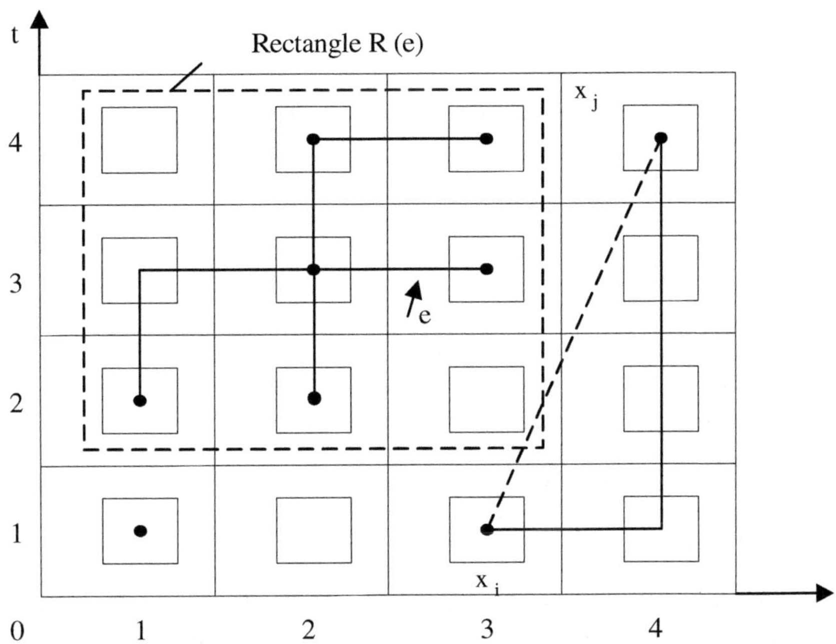

Fig. 2.21. Model of a plane for the placement of graph nodes

Into the initial plane, the Cartesian system of coordinates with axes s and t is introduced, divided into rectangular cells. In each cell, a node of the graph or hypergraph can be placed. The distance between nodes is calculated on the basis of one of the known formulas [41]:

$$d_{i,j} = |s_i - s_j| + |t_i - t_j|, \tag{2.39}$$

or

$$d_{i,j} = \sqrt{|s_i - s_j|^2 + |t_i - t_j|^2}\ . \tag{2.40}$$

Here (s_i, t_i); (s_j, t_j) are the coordinates of x_i, x_j, the elements (graph nodes), $d_{i,j}$ is the distance between elements x_i, x_j on the plane. In addition, expression (2.39) allows for defining the Manhattan distance, i.e. the distance between two nodes determined on vertical and horizontal directions. The step (distance) between two adjoining cells (nodes) along the horizontal or vertical axis is equal 1. Expression (2.40) allows for calculation of the rectilinear distance between two nodes. In the case of an edge of a hypergraph its length is taken as half of the perimeter of the rectangle covering its trailer points (the length of edge e in Fig.2.21 is equal to 6). Then the total length of all edges of a graph model is defined by the known formula [41]:

$$L(G) = \frac{1}{2} \sum_{i=1}^{n} \sum_{j=1}^{n} d_{i,j} c_{i,j}. \tag{2.41}$$

Here L(G) is the total length of edges of a graph, $c_{i,j}$ is the number of edges connecting nodes x_i and x_j. The number of crossings of the graph edges can be defined by the formula

$$S(G) = \frac{1}{2} \sum_{u_{i,j} \in U}^{n} S(u_{i,j}). \tag{2.42}$$

Here $S(G)$ is the total number of crossings of edges of the graph, $S(u_{i,j})$ is the number of crossings of an edge $u_{i,j}$, connecting nodes x_i, x_j.

Let us consider the algorithms of minimization of L(G) and S(G) (L(G) → min, S(G) → min). Note that minimization of the total length of graph edges usually results in minimization of the number of crossings. Such simultaneous minimization of length and number of crossings takes place up to some limit, beyond which reduction of length entails an increase of the number of crossings.

In OT of placement the input data are the set of nodes of a graph or a hypergraph, X = {x_1, x_2 …, x_n}, set of edges of the graph, U = {u_1, u_2…u_m}, or of hyper graph E = {e_1, e_2 …, e_m}, and set of locations (places) Q = {q_1, q_2 …, q_f}. In addition, |X| ≤ |Q|, if placement of nodes in definite locations is carried out. Sometimes the problem is formulated as the placement of edges and then the number of locations should be ≥ than the number edges placed.

Let us formulate the problem of placement as assignment of each node of a graph (hypergraph) to a unique location so as to optimize TF. The expression (2.41) defines the TF. The target data will be the locations assigned to each node. The following conditions have to be observed as constraints: For every $x_i \in X$ only one location $q_i \in Q$ is selected. Accordingly, to each location $q_i \in Q$ at most one element $x_i \in X$ is assigned. Thus, the OT of graph node placement is formulated. It is known [41, 84], that the placement problem is an NP-complete problem. In the search for heuristic procedures securing solution quality and reducing computational complexity, we look for algorithms with ART of O(n), O(nlogn), O(n^2).

Consider placement of graph nodes on a plane in a lattice of a given configuration. Two approaches with application of ES and GA to such a placement problem shall be outlined. The first one uses the Cling methodology [90, 91] for placement of elements, identical along height and different with respect to width, using ES. The second approach is based on the ideas of Paris, Mailler, Shervani and Masumder [84, 93, 94].

The first approach works as follows. The initial population of solutions is usually formed in a random manner. Then, to each element of the population OM is applied, TF is calculated, solutions are sorted and appropriateness of placing elements in given positions is assessed. In case of necessity the procedure returns to the mutation block and the algorithm proceeds similarly until (premature) convergence or execution of a predefined number of iterations. This approach is simple to implement and has linear ART, but the obtained local optima can be far from global. This results from very rough simplification of the ES, which does not use all the mechanisms of evolution.

A modification of this approach for placement of graph (or hypergraph) nodes on a plane or on a line, with optimization of expression 3.15 is suggested. It consists in the use of additional GO, and of the homeostatic, synergetic and hierarchical principles of management, in analogy to the NS. In Fig. 2.22 the scheme of the modified algorithm of placement of the graph nodes on a plane is given, based on models of evolution [95]. Let us note that the developed architecture of search involves parallel work of algorithms and joint application of heuristic, stochastic and genetic search.

In the GA of Fig. 2.22 the initial population is designed in three ways, A_1, A_2, A_3. In case A_1 this population is formed randomly. In case A_2 the initial population results from a number of applications of the consecutive algorithm. In case A_3 the population is generated by a joint action of the random and directed procedure. Let us note that the DMP, using the block of evolutionary adaptation, can create any number of populations by other methods.

After formation of the initial population, the mutation operator is applied to its each element, and it can be done independently for each element of the population. Then, an exchange of alternative solutions between populations of solutions is possible. The exchange takes place on the iteration, when premature convergence occurs, i.e. the chromosomes with the best values of TF do not appear.

Then, the operator of inversion (IO) is applied. Two versions of IO are proposed: random (IO_1) and directed (IO_2). Then, three versions of the segregation operator (SO) are applied. In CO_1 the "greedy" strategy is used. The strategy of gold section is used in SO_2. The Monte Carlo method is used in SO_3. Then the algorithm passes to the execution of the translocation operator (TO). Two basic versions of TO are applied, as well, the random (TO_1) and the directed (TO_2). As the last GO the CO is carried out. Here CO_1 is a single-point CO, CO_2: point-to-point CO, CO_3: ordered CO, CO_4: cyclic CO, CO_5: modified CO. Yet, in general, the actual sequence of execution of the particular GO depends on the external environment and the BEA block.

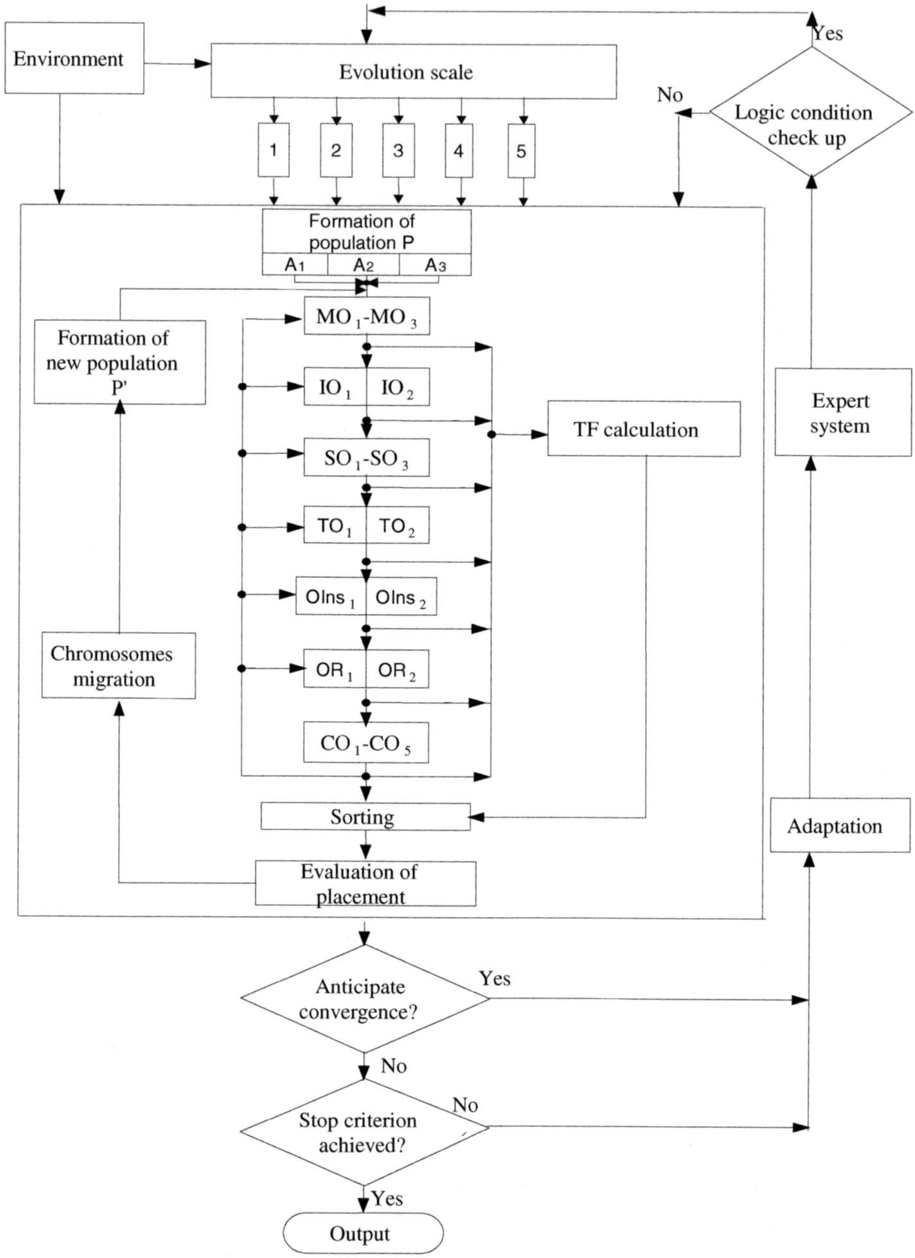

Fig. 2.22. The scheme of the algorithm of placement based on evolution models of

The estimates are calculated for each node of the absolutely minimal placement for the given configuration on the plane. After the execution of all the GO the results are compared with the estimated ones. Upon obtaining the desired results the algorithm terminates its functioning. TF value is calculated for each chromosome after application of CO and the solution variants are selected on its basis. Then the new population is generated, and the process proceeds similarly. The exit from each operator is possible separately upon reaching a local optimum.

Let us consider an example of GA for placement of graph nodes on a line. Let the graph shown in Fig. 2.23 be given. The task consists in placement of the graph nodes on a line with six locations, so as to minimize the total length of edges. Let for the graph G = (X,U), |X| = 6, |U| = 10, shown in Fig. 2.23, the following initial population of solutions (accommodation of nodes the graph in a ruler) be randomly obtained:

P_1: 1 2 3 4 5 6, P_3: 2 4 6 1 3 5, P_5: 4 6 2 5 1 3,

P_2: 1 3 5 2 4 6, P_4: 3 5 1 6 4 2, P_6: 5 1 4 3 6 2.

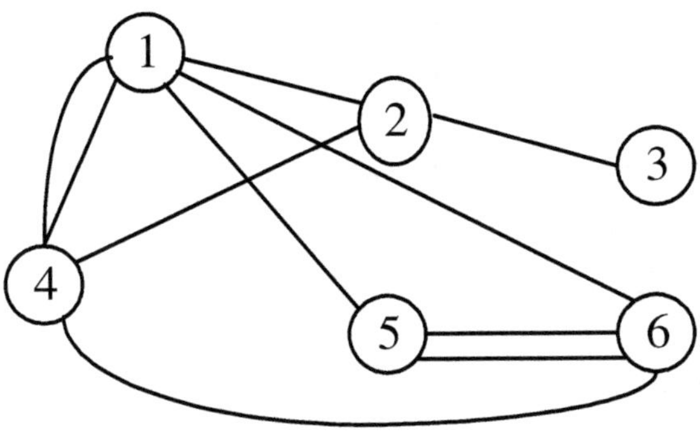

Fig. 2.23. Graph G

According to [41], we will determine an estimate of the minimal total length of edges for this graph. It is defined for the arrangement of graph edges in the shortest distances, not accounting for the isomorphism. Thus, multiple edges are ordered according to their length and assigned to the nearest positions in a decreasing order. In our case for the graph G from Fig. 2.23 the estimate will be equal $L\ (G\Delta)_{min} = 13$. We will calculate now the TF value for all chromosomes in the population. Then, the total length of connections for the particular solutions, P_1 through P_6, will be:

$L(G)_1 = 1{\cdot}4 + 2{\cdot}2 + 3{\cdot}2 + 4{\cdot}1 + 5{\cdot}1 = 23;$

$L(G)_2 = 28;\ L\ (G)_3 = 22;\ L\ (G)_4 = 19;\ L\ (G)_5 = 24;\ L\ (G)_6 = 25.$

The ordering of the chromosomes according to their TF is, then: P4; P3; P1; P5; P6; P2. Starting from P_4, we apply MO1. Let mutation point be defined randomly between elements 5 and 6. Then we get

P_4: 3 5 1 6 4 | 2 L $(G)_4 = 23$,

P'_4: 3 5 1 6 2 4 L $(G')_4' = 20$.

Two basic solution procedures are now proposed. In the first one mutation is carried out for all P_i from population P and the new population P' is thus constructed. In the second proposal the population changes after every application of the MO. For example, we introduce P'_4 into P' and exclude P_2, so that P' = {P_4, P'_4, P_3, P_5, P_1, P_6}. Let us show now the application of the operator of inversion IO1. Let us take P4, in which, randomly, the point of inversion is determined elements 1 and 2. Then

P_4: 3 | 5 1 6 4 2 L $(G)_4 = 23$,

P'4: 3 | 2 4 6 1 5 L $(G)_4' = 16$.

The value of L(G) approaches the estimated global minimum. Basically, at this point the process of placement could terminate. Fig. 2.24 shows this placement:

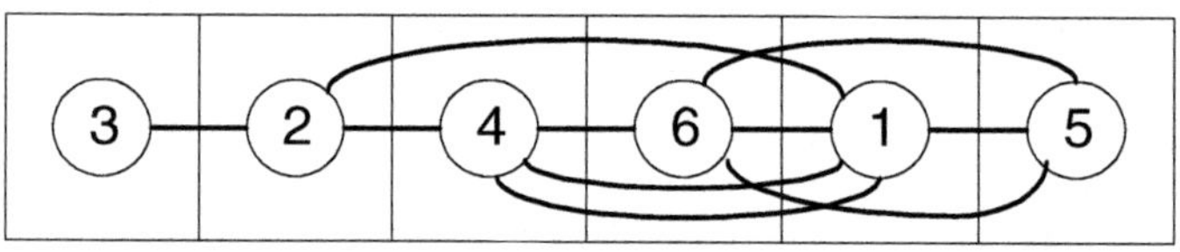

Fig. 2.24. Placement of graph G (L (G) = 16)

Let us note that it is possible in this algorithm to implement feedback for preservation of homeostasis, based on homeostatic management (HM), after execution of each GO. So, for example, for the best chromosome P ' 4 the directed mutation operator (OM2) can be carried out:

P'4: 3 2 4 6 | 1 5 L $(G)_4' = 16$,

P''4: 3 2 4 1 | 6 5 L $(G)_4' = 13 = L\,(G)_{min}$.

In Fig. 2.25, the placement of the graph G, producing the global minimum value of L(G) on a line, is shown. It is possible to obtain such placement in all kinds of linear settings by applying the GA for partitioning, described above. For this purpose it is necessary to break the graphs into a given number of positions, with minimization of the number of edges between them, and then to perform optimization of placement inside each linear setting, using the methods of genetic search.

A rough expression for the computational effort of the here presented linear algorithm is:

$$T \approx [N_p t_{p+} N_p\,(t_{mo} + t_{io} + t_{so} + t_{to} + t_{CO})]\, N_G, \quad (2.43)$$

where t_{mo} is computational effort of the mutation operator; t_{io} is computational effort of the inversion operator, t_{so} — is the computational effort of the segregation

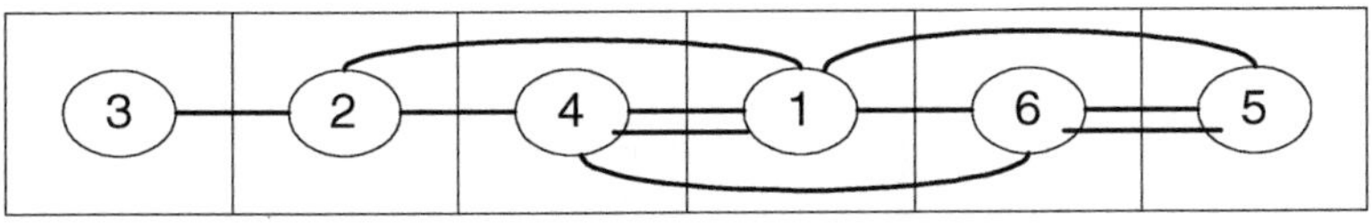

Fig. 2.25. Optimum placement of graph G in a linear setting

operator; and t_{to} – of the translocation operator. ART of linear placement varies from O (n) to O (n^2).

The second approach to placement of graph nodes on a plane is based on the use of classical GA and the principles of meta-genetic optimization. It allows for increasing the quality of solutions at the cost of increased ART.

Let us describe the application of micro-, macro- and meta-genetic optimization and genetic search to placement of graph nodes in a lattice. The block diagram of the algorithm featuring relative balance is shown in Fig. 2.26. Here, population is generated in three ways: deterministic, P_1, random-deterministic, P_2, and random, P_3. The minimal size of the population is n/2, where n = |X|. The minimal number of generations is N_G = 10n. The scheme of the algorithm consists of two BB. It is constructed in such a manner that during its functioning a relative balance is established. In addition, at the input to one BB a set of randomly obtained chromosomes is provided, and at the input to the second a predefined set chromosomes. Then, we simulate the development of chromosomes by running GA with CO, MO, SO, TO, IO, and other GO, developed by the authors. Upon calculation of TF, we find the best placement with the given parameters. We carry out the algorithm by selecting the parameters and using BEA. The solutions with best TF values are selected from the set of parents and descendants, and copied in a new population. Population size remains constant.

The choice of the best set of parameters for the final population is made finally. Population size is usually determined on the basis of DMP's experience and available computing resources. Let us note that the global minimum can both be obtained at the first generation and not found at the last one. The application of the standard GA and SGA can lead to non-realistic placements, and so it is necessary to use various modified GO and a new architecture of GS implementation.

Fig. 2.27a shows a graph G = (X, U), |X | = 9, located in a 3x3 lattice, with L(G) = 30, expressing the TF. We wish to pass to an order, with L(G) → min, from the initial chaos (random arrangement of nodes). Let us construct a FS. We shall present each line of the lattice as one node, and a new graph G_1 with $|X_1|$ = 3, is produced for this purpose. In Fig. 2.27b the result of such a transformation is shown. Let us apply the algorithm of aggregation of fractals. Fig. 2.27c shows the optimum result for the arrangement of the graph of Fig. 2.27b in a linear setting. After the transition from the fractal graph G1 to the graph G we shall obtain the arrangement of the graph G in the lattice as shown in Fig. 2.27d. here, L(G) = 25. Similar transformations for the graph G (Fig. 2.27d) are feasible with construction of fractals on graph G_2 with $|X_2|$ = 3. These transformations are shown in

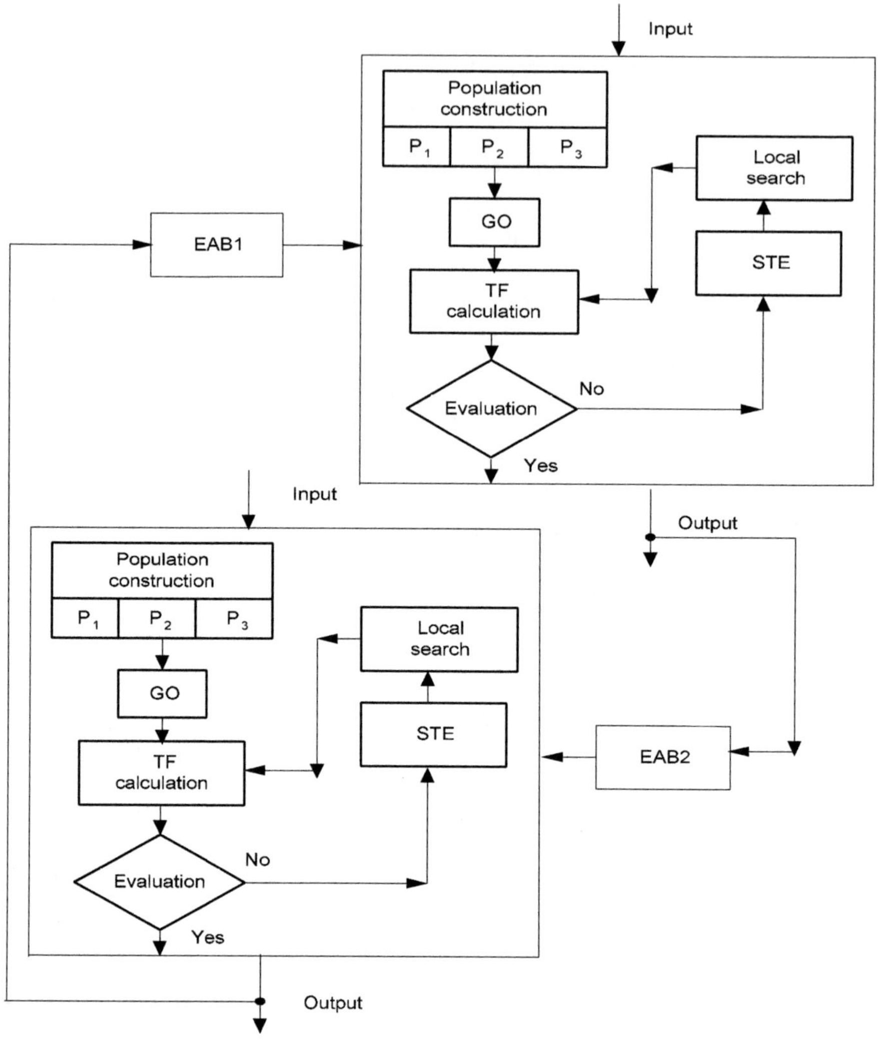

Fig. 2.26. Block diagram of the algorithm with relative balance

Figs. 2.27e and f. The final placement of the graph G of Fig. 2.27a, with L(G) = min = 20 is given in Fig. 2.27g. Let us note that it is possible to carry out such transformations for a graph placed in a lattice, not only on lines and parts of a lattice, but also for an area of any configuration.

In the summary of this subsection we shall note that the complexity of the described algorithms of placement is polynomial. For one generation ART is contained between O (n^2) and O (n^3).

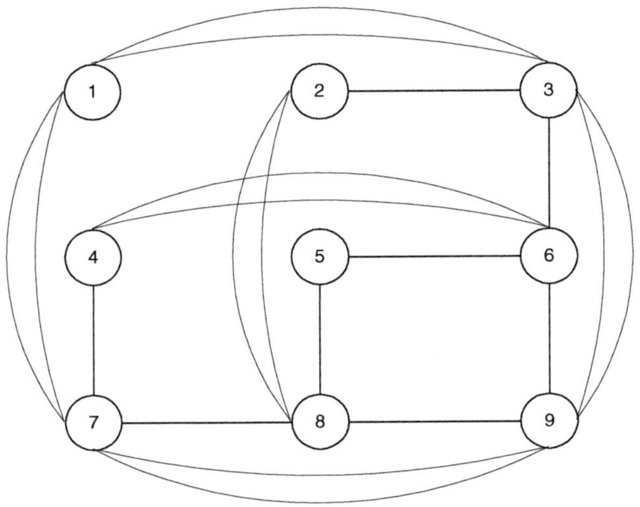

Fig. 2.27. a. Graph G = (X, U), |X| = 9

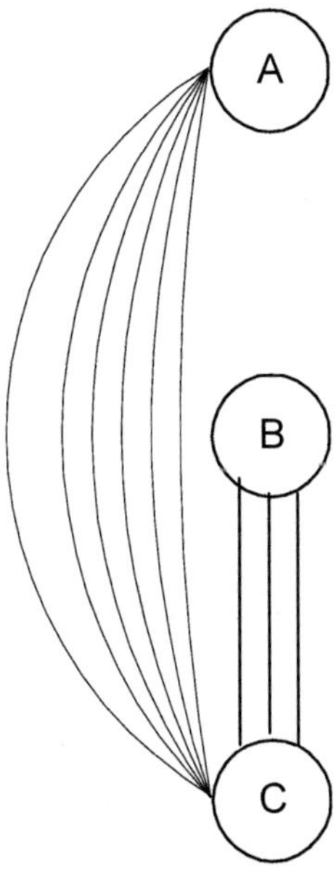

Fig. 2.27. b. Linear graph G1 before transformation

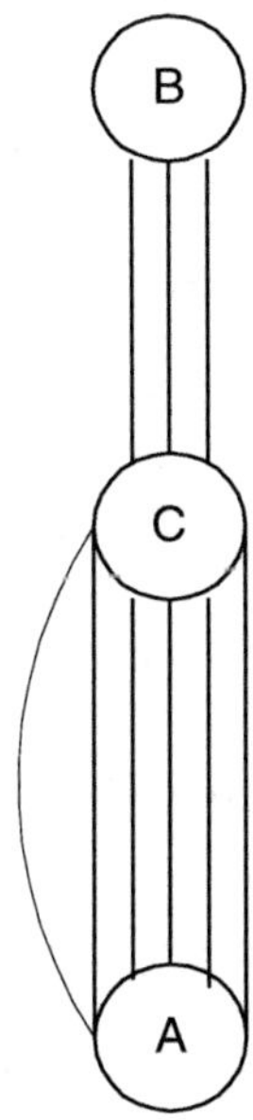

Fig. 2.27. c. Linear graph G1 after transformation

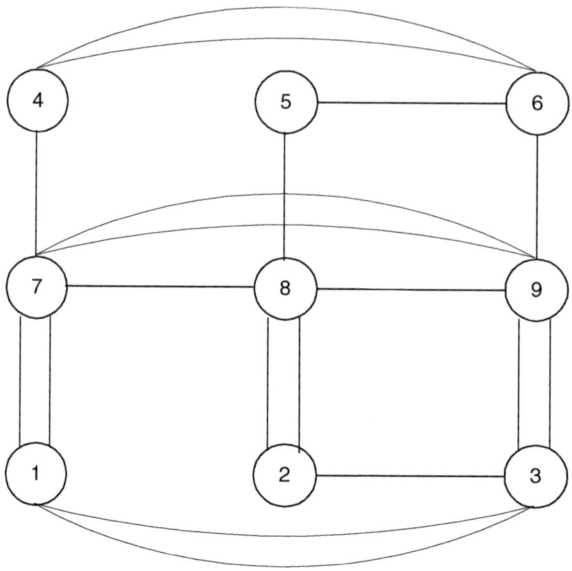

Fig. 2.27. d. Graph G after the first transformation

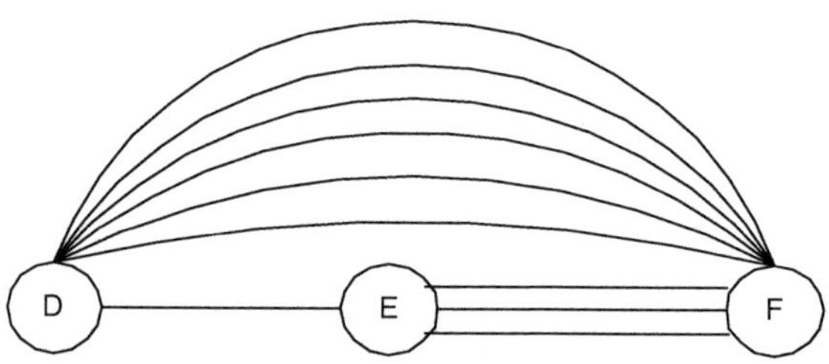

Fig. 2.27. e. Linear graph G2 before transformation

2.3.3 Solving the Traveling Salesman Problem by Simulated Evolution

Determining the methods of solving the Traveling Salesman Problem (TSP further on) is one of the major optimization tasks in science and practice. Let there be given a graph G = (X, U), where X, |X| = n, is the set of nodes (cities), and U, |U| = m, is the set of edges (possible routes between cities). A matrix of numbers R(i,j), where i, j ∈ 1, 2, …, n, representing the cost of moving from x_i to x_j is given. The problem is to find a rearrangement of the elements of X, such that [55, 96—99]:

$$TF(\varphi) \equiv R(\varphi(1), \varphi(n)) + \Sigma\{R(\varphi(i), \varphi(i+1))\} \rightarrow \min. \quad (2.44)$$

It corresponds to finding a route with minimum cost leaving the node 1, passing all graph nodes and getting the node n. Let us write down, for example, the matrix

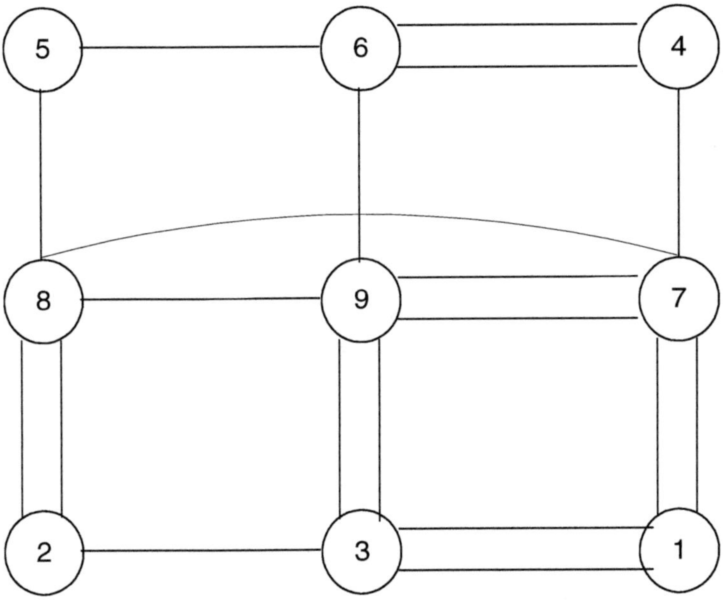

Fig. 2.27. g. Final placement of graph G

of contiguity of the graph G = (X, U), where |X| = 7, |U| = 21, as shown in Fig. 2.28. For the route φ(1) = <1,2,3,4,5,6,7> (chromosome P_1), we have TF(φ(1)) = 1 + 1 + 1 + 1 + 1 + 1 + 1 = 7, and for the route φ(2) = <7,2,5,4,3,6,1> (chromosome P_2), we have TF(φ(2)) = 3 + 6 + 1 + 1 + 5 + 3 + 1 = 20. Hence, route φ(1) is preferred over φ(2) in terms of TF value. We shall note that the value of TF(φ) does not depend in this specific case on the choice of the initial node – the beginning of a route. For example, TF(φ(3)) = TF(φ(2)) = 20, where φ(3) = <2, 3, 4, 5, 6, 7, 1> (chromosome P_3).

R =

	1	2	3	4	5	6	7
1	0	1	2	7	6	3	1
2	1	0	1	3	6	5	3
3	2	1	0	1	3	5	6
4	7	3	1	0	1	3	6
5	6	6	3	1	0	1	4
6	3	5	5	3	1	0	1
7	1	3	6	6	4	1	0

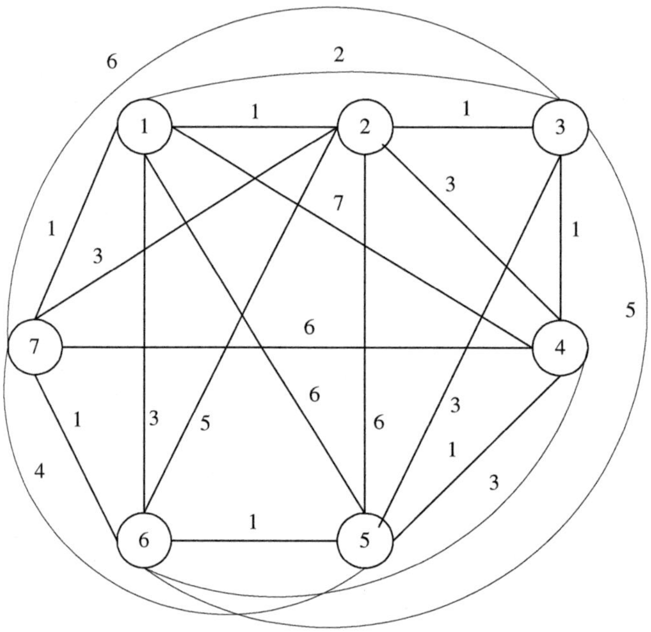

Fig. 2.28. Graph G = (X, U)

Let us consider an evolutionary strategy, which is a combination of GA, evolutionary programming, models of evolutions, and also the local search methods. We shall, first, describe the characteristics of the genetic operators used to solve the TSP.

There exists, namely, for the TSP, a specially developed GO, meant to guarantee obtaining feasible descendants. We propose an algorithm of construction of the modified translocation operator (OT) for the TSP, based on a greedy strategy:

1. For each pair of chromosomes select randomly a break point and as the number of the initial node, take the number of the marked gene in a chromosome.
2. Compare the partial cost of routes defined by the current nodes analyzed, and select among them the route with the least cost.
3. If the thus chosen node has already been included in the partial route, take randomly a node from among the yet not marked ones. Update the TF value according to the new node.
4. Upon the premature formation of a cycle find another shortest route out.
5. Try inverting the elements of the entire route or its parts and select the best variant.
6. Repeat steps 2 and 3 until a Hamiltonian cycle with minimal total cost of edges is constructed.
7. Terminate the algorithm.

In this greedy approach the problem-related knowledge is used namely that the choice of the shortest route is usually preferable. However, similar simple assumptions do not always result in an optimum, as sometimes it is required to neglect some local prize for the sake of achievement of the global purpose. The greedy approach, based on problem-related knowledge, improves the speed of convergence and eliminates all the wrong changes obtained from the CO. When a deadlock is encountered on a route, the use of a combination of greedy strategy and "horizon" method is proposed.

Let us describe the modified combined CO:

1. Randomly select the starting node in the first parent.
2. The current parent becomes first parent.
3. Definition of the nodes in the combined CO. Select one of the neighboring nodes (nodes located in the chromosome immediately to the right and to the left of the considered node) in the analyzed parent. The nearest of these nodes, which has not been yet incorporated in the constructed TSP route, becomes the new current node.
4. Analyze the branches of the decision tree for the given depth of search.
5. Repeat step 3 until all nodes will have been visited. CO is formed as a sequence of visited nodes.
6. Termination of the algorithm.

For example, let a graph G = (X, U), be given as in Fig. 2.29, with X, |X| = 9, the set of nodes (cities), U, |U| = 12, the set of edges (possible routes between cities with given cost). We want to find a route (or routes) with minimum cost from node A to B. Let us construct a population of the alternative candidate solutions $P = \{P_1, P_2, P_3, P_4\}$:

P_1: A 1 2 3 B

P_2: A 5 4 3 B

P_3: A 1 6 7 B

P_4: A 5 2 7 B.

As the result of functioning of the modified greedy CO we obtain the descendant chromosome $P_5 = P_3$: A 167 B (Fig. 2.30). Thus, L(G) = 13. Let us note that a single application of the greedy GO allowed for finding the shortest way from node A to B. We use a combination of the greedy strategy and the method of "horizon" of depth 2. We analyze the previous population. After the first step the node 1 and the partial routes A-1-6 (L(G) = 11) and A-5-2 (L(G) = 6) are preferred. However, on the second step the preference goes back to the partial route A-1-2 (L(G) = 11). By continuing the algorithm, we obtain the globally optimum solution P_6: A 523 B (L(G) = 8) (Fig. 2.31).

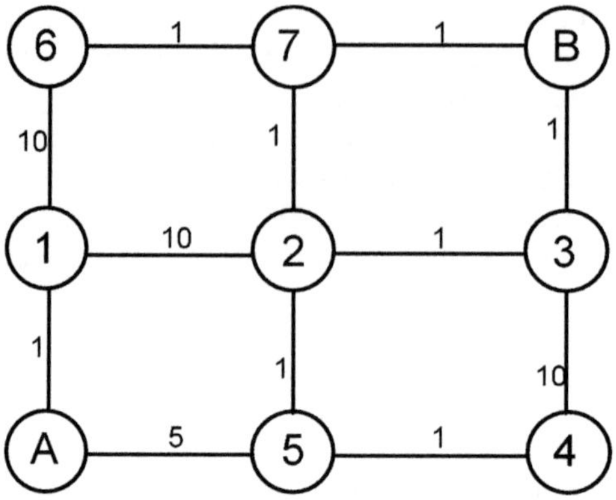

Fig. 2.29. Graph G

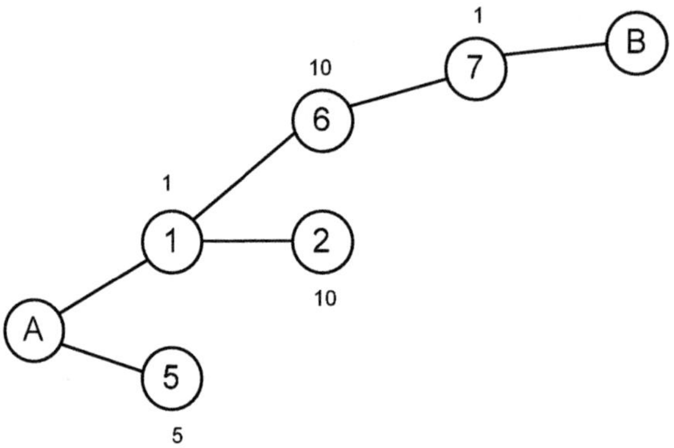

Fig. 2.30. Result of the greedy CO application

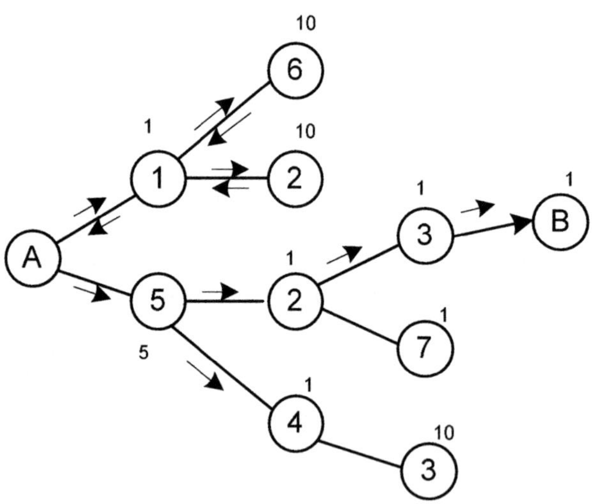

Fig. 2.31. Result of the combined strategy application

Let us consider the strategy of finding alternative solutions to the modified TSP, when the choice of the first node is insignificant. Let the graph G = (X,U), where X, |X| = 4 is the set of nodes, U, |U| = 6 is the set of edges (possible routes between cities with a given cost) be given as a contiguity matrix.

R =

	1	2	3	4
1	0	2	3	4
2	1	0	2	3
3	1	2	0	5
4	3	7	6	0

Let us generate in a random manner the population of alternative solutions:

$P^0 = \{P_1, P_2\}$, $P_1 =$ 4 2 3 1 4 (TF = 7 + 2 + 1 + 4 = 14); P_2 = 1 4 3 2 1 (TF = 4 + 6 + 2 + 1 = 13), so that TF_{avg} = 13.5. The principle of realization of the combined strategy is shown in Fig. 2.32. First, we select node 4 and apply the greedy CO. We obtain the alternative solution P_3 = 4 3 1 2 4 with TF = 6 + 1 + 2 + 3 = 12. Then, we select node 2 and proceed similarly. The result of the second step is shown in Fig. 2.33 (P_4 = 2 1 4 3 2 with TF = 1 + 4 + 6 + 2 = 13).

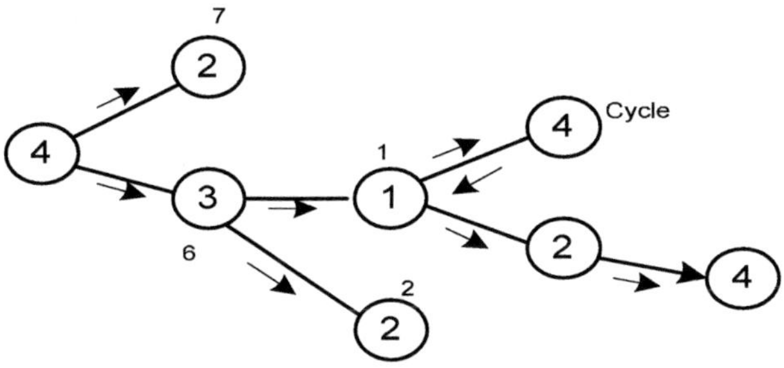

Fig. 2.32. Realization of the greedy CO for the case of the graph represented by the contiguity matrix R above

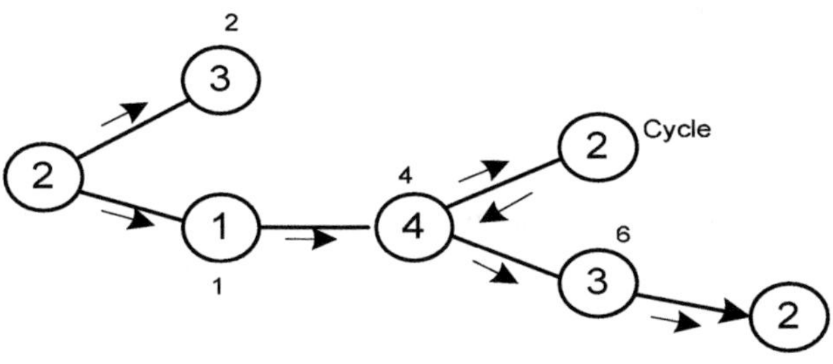

Fig. 2.33. Result of the second step

Let us construct a new population. We leave in the new population three chromosomes, $P_1 = \{P_2, P_3, P_4\}$. By applying the same algorithm, we select node 3. We obtain $P_5 = 3\ 1\ 2\ 4\ 3$ with TF = 1 + 2 + 3 + 6 = 12 (Fig. 2.34). Then, we select node 1 and proceed similarly. The result of the second step for the population P_1 is shown in Fig. 2.35 ($P_6 = 1\ 2\ 4\ 3\ 1$ with TF = 2 + 3 + 6 + 1 = 12). Let us apply MO to chromosome P_3, having chosen mutation point between the second and third gene. We obtain the alternative solution $P_7 = 4\ 1\ 3\ 2\ 4$ with TF = 11. Let us apply MO to chromosome P_5, having chosen mutation point between the first and second gene. We obtain the alternative solution $P_8 = 4\ 1\ 3\ 2\ 4$ with TF = 11. So, the alternative solutions $P_7 = 3\ 4\ 1\ 2\ 4$ and $P_8 = 3\ 1\ 2\ 4\ 3$ are the best.

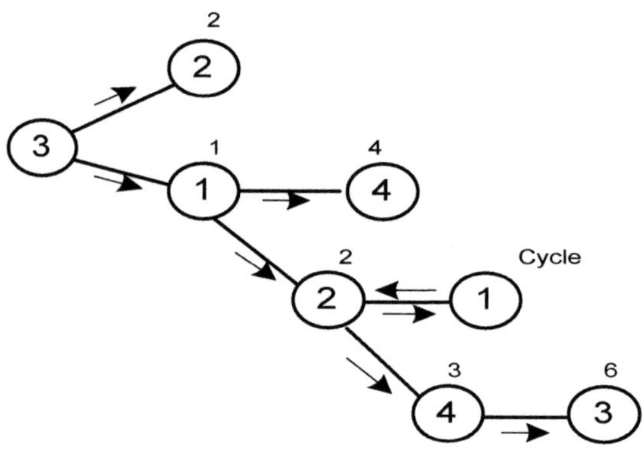

Fig. 2.34. Alternative solution P_5

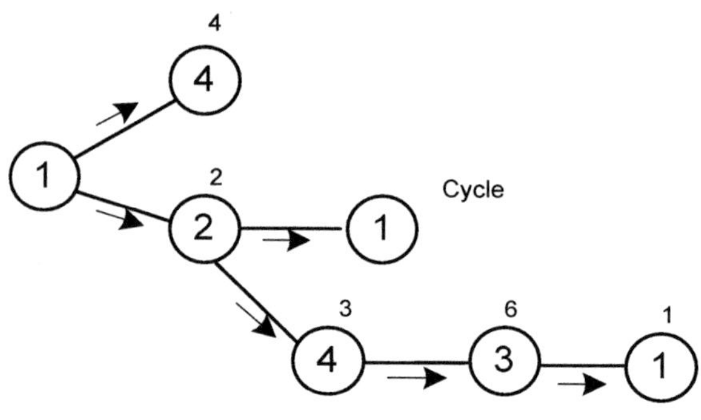

Fig. 2.35. Alternative solution P_6

The architecture of genetic search for the solution to the TSP is similar to the scheme of the algorithm for the placement problem. In distinction from the latter it is proposed to introduce a block of local search, allowing for finding local optimum of the TSP. All genetic operators are focused on the use of knowledge of the problem to be solved. We propose to change the order of application of the genetic operators. In solving the TSP, the main role is played by the CO and its updating.

Hence, combined CO is tried out first in the proposed scheme. Application of other genetic operators will depend on the concrete features of the given TSP. The reduction block changes population size. The following heuristics is proposed for the case of hitting a local optimum:

- to change the depth of the search "horizon";
- preservation of one chromosome: of all routes having identical length, only one remains unchanged; for all the remaining ones the modified genetic operators are carried out;
- use of various feedbacks to change the parameters of genetic search.

Let us note that such genetic operators and heuristics do not function all the time, but only in situations, when the monotony in a population attains a definite high level. The use of joint models of evolution and external environment regulates and controls this process of genetic search.

The results obtained from solving of the TSP with genetic algorithms allow for drawing conclusions on the efficiency of the method and convenience of use of similar approaches in solving other optimization problems. In summary we shall note that the running times of the described algorithms are polynomial and are contained between O (n) and O (n^3).

2.3.4 Greedy Heuristics for Solving the Problems of Coloring, Isomorphism, Construction of Cliques and Independent Sets of Graphs

Let us consider the problem of graph coloring (GC). Coloring of the nodes of a graph G = (X, U) [41, 66] is equivalent to partitioning of the set of nodes X into L not overlapping classes (subsets), i.e.,

$$X_1, X_2 \ldots X_L;\ X = \bigcup_{i=1}^{L} X_i\ ;\ X_i \bigcap X_j = \varnothing\ ;\ i,j \in I = \{1,2 \ldots L\}, \tag{2.45}$$

such that no subset X_i contains any adjacent nodes. The name of the problem refers to assignment of various "colors" to individual classes, so that no two adjacent nodes have the same color. In the case of (2.45) the graph is referred to as L-colored.

The consecutive and greedy heuristics (GH) are the basic strategy for GC, yielding a local optimum at the first attempt.

The consecutive heuristics consists in the following. First, a list is made ordered according to the decreasing degrees of nodes. The first node is removed from the list and colored with color 1. The list is inspected and all nodes not adjacent to the removed ones, are colored 1 and are themselves removed from it. The second node on the list (if still on it) is colored 2 and removed. The algorithm proceeds similarly, until all nodes are colored.

The greedy algorithm is the optimization algorithm, which is based on GH. It looks through the set of alternatives and chooses the first-best solution. Let us write down of the basic greedy heuristics for coloring [66].

Heuristics 1. "The first are the biggest". The graph nodes are ordered according to their decreasing local degrees. Then, in this order, each node, after the analysis of connections, is colored with the least possible color number.

Heuristics 2. "The least are the last". The node with the smallest local degree is selected and removed from the graph together with the incident edges. The procedure is repeated for the remaining graph. The nodes of the graph are colored with colors of the least possible numbers in the order opposite to the order of removal.

Heuristics 3. "Partitioning according to the contiguity". At the beginning of the algorithm the node with the maximum degree is selected and colored with color 1. Then, all the not colored nodes are divided into two subsets: X_1, containing nodes adjacent to the colored one, and X_2, containing the remaining nodes. Next node for coloring with the least possible color number is the one with the biggest degree in X_1. The process continues similarly until coloring with color 1 is terminated. After this, the nodes colored by color 1, are removed and the procedure continues to complete the GC.

Heuristics 4. "Coloring of the ordered nodes". For node x_i color k is selected, if and only if among the nodes, adjacent to x_i the nodes exist already colored with colors c = 1,2... k-1, k + 1, k + 2, ..., and there are no nodes colored with color *k*.

Heuristics 5. "Coloring with ordered colors". Initially, the set of colors is given for each node, with which the nodes can be colored. The nodes are ordered in the following manner. For each node the number $N = \rho(x_i) + z_i/C$, where $r(x_i)$ is the local degree of node x_i, z_i is the number of the forbidden colors for the given node, C is the total number of colors, is determined. The node x_i with $\min_i N$ is removed from the graph together with the adjacent edges. The procedure is then repeated, etc. The nodes are colored with the color of the least possible number in the order opposite to the order of removal.

Let us write down a simple algorithm based on one of the outlined GH:

1. Order the nodes according to decreasing values of the local degrees.
2. Assign the first available color to each node on the ordered list, if possible.
3. Continue with step 2 to complete GC.
4. Terminate the algorithm.

The algorithm, developed on the basis of such heuristics is greedy, since it colors the graph consistently node after node, assigning a color to each node, if it is feasible, without considering the global consequences of such coloring. The greedy algorithm of GC is very fast (ART at $O(\alpha n)$), because it considers just one of possible variants of coloring each node.

Let us consider, for example, GC for the graph of Fig. 2.36. The heuristics develops the following order: <9 7 1 8 6 3 2 5 4>. Node 9 is colored first, the algorithm leaves node 7 not colored, since it is connected with 9, and node 9 is already colored. The heuristics does not color node 8 for the same reason. Node 1 is then

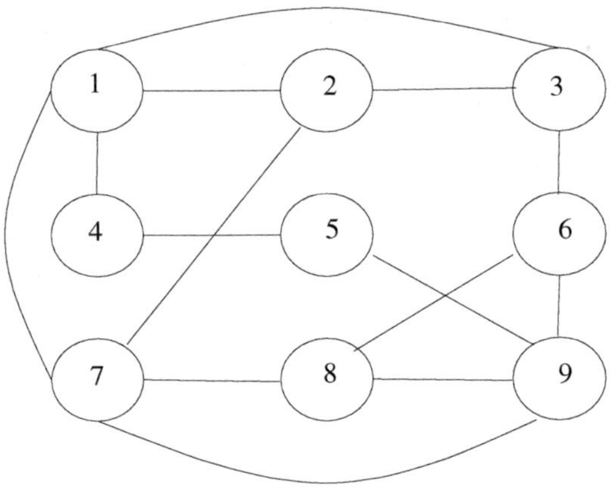

Fig. 2.36. Graph G

colored and other ones are not colored. Hence, $X_1 = \{9, 1\}$. On the second step we obtain $X_2 = \{7, 6, 5\}$. Then, $X_3 = \{8, 3, 4\}$ and $X_4 = \{2\}$. Consequently, graph G of Fig. 2.36 is colored with 4 colors.

Consider now the graph of Fig. 2.37. This wheel W_7 contains seven nodes. The first six nodes form a ring, and the seventh node is located at the center of the ring and is connected to all other nodes. The GH creates the following order: <7 6 5 4 3 2 1>. The algorithm will color the nodes as follows: $X_1 = \{7\}$; $X_2 = \{6, 4, 2\}$ and $X_3 = \{5, 3, 1\}$. So, the wheel W_7 of Fig. 2.37 is colored with three colors. Let us note that if in a graph, similar to a wheel, there is one node, adjacent to all the other ones, it is colored with one color and removed from the list.

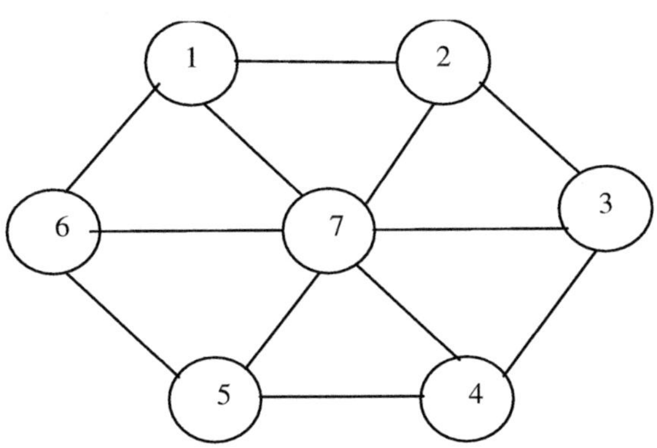

Fig. 2.37. Wheel W_7

Since the greedy and consecutive heuristics for GC work fast, it is proposed to combine them with GA in the framework of the concept developed by the authors.

The strategy of overlapping assumes that in the algorithm developed the coding meant for hybrid GA is used. The coding used in GH is based on regularization of the graph nodes. It orders nodes according to values of their local degrees and then decodes this sequence, assigning to each node in the order the first available color, where availability is based on the use of colors for previous nodes. Such a way of coding in solving the GC problem is based on ordering of nodes of the graph in a certain way, and then decoding them according to GH.

First, it is necessary for the GA to replace the binary TF with TF for node coloring. In constructing the TF the first gene (node) is chosen in a chromosome and the first available color is assigned to it. At each step the previously used colors are accounted for. The rearrangement of the list of elements produces the regularization of the elements on the list (a trivial rearrangement simply leaves the previous order). The following step in GA is to replace the used representation by the one of the changed list. This means that each chromosome is a rearrangement of the list of graph nodes. The GH sorts the list of nodes and transfers the result to the decoding heuristics.

The following step in the GA is the replacement of the random or problem-related-knowledge-based initialization though rearrangement. In GA we generate the rearranged list of nodes randomly and in a definite way. GH generates only one chromosome, but this chromosome has a chance to be better than on the average. Such procedure guarantees that with the help of GA we shall at least not worsen the results of GH, as with the elite and other procedures of selection used for generating initial population. We shall obtain the changed list of nodes of the graph as the result of changes of elements on the ordered list. The ordered list and the list with changes shall be used in further procedure. Upon application of the GO to the obtained chromosome, we determine TF. In the algorithm considered the elite selection and GO, described above, are applied.

The process of overlapping is linked with changes in the reproduction module, where GO, working with ordered representations, are used.

Let us consider the example of graph from Fig. 2.36. Let alternative solutions be coded as follows:

P_1: 1 2 3 4 5 6 7 8 9

P_2: 9 7 5 3 1 8 6 4 2

BS:0 1 0 0 0 1 0 0 0.

After step two we obtain:

P_1: – 2 – – – 6 – – –

P_2: 9 – 5 3 1 – 6 4 2.

And the final result:

P_3: 9 2 5 3 1 6 7 4 8

P_4: 9 8 5 3 1 7 6 4 2.

Here, BS is the binary string showing contiguity of the appropriate nodes in P_1 and P_2. So, 1 in BS shows that nodes 5 and 9 (Fig.2.36) are connected. If adjacent nodes in the given parents are not present, then BS is filled with 0's and 1's randomly. Having executed greedy decoding, we see that chromosome P_3 corresponds to the following GC: $X_1 = \{9,2,4\}$, $X_2 = \{5,3,7\}$, $X_3 = \{1,6\}$ and $X_4 = \{8\}$. For P_4 we get as GC: $X_1 = \{9,3,4\}$, $X_2 = \{8,5,1\}$, $X_3 = \{7,6\}$ and $X_4 = \{2\}$.

As it is easily seen, a part of structure of each parent is fixed by one parent (as specified in a bit line). The other elements are re-ordered so that the elements remain in the same order, in which they were in the other parent. This operation has running time of $O(n^2)$, where n is the size of the chromosome, a factor of inclination of a parabola.

We propose a new GS with simultaneous decoding of chromosomes. Let for the graph G of Fig. 2.37 a population $P = \{P_1, P_2, P_3, P_4\}$ be given:

P_1: 1 2 3 4 5 6 7 8 9,

P_2: 9 8 7 6 5 4 3 2 1,

P_3: 1 3 5 7 9 2 4 6 8,

P_4: 9 7 5 3 1 8 6 4 2.

In chromosome P_1 the gene corresponding to the node with maximum degree is selected. If there are more such nodes, they are analyzed consistently in any arbitrary order.

We select element 9 in the first chromosome. According to the greedy crossover operator (GCO), node 1 is selected as the next one. Further GCO steps for the entire population result in impasses. Hence, first color, $X_1 = \{9,1\}$ is determined. The elements 9, 1 are excluded from the population P. Then, element 7 is selected from P_1. We obtain the second color, $X_2 = \{7, 6, 4\}$ through application of the "greedy" CO. Ultimately, we find in a similar way the GC (Fig. 2.36) consisting of four colors, with $X_3 = \{8, 2, 5\}$ and $X_4 = \{3\}$. The technique allows for reducing the chromosome (alternative solution) size at each step and so also the time of GC.

The structure of the hybrid GA, used for GC, is shown in Fig. 2.38. In this scheme the ideas of joint evolution, establishment of balance, adaptation by external environment, and active interaction are used. The reference to the external environment, hierarchical management of GS, local search for solutions and application of all the modified GO form the basis of GS and the search methods developed by the author. The ART for the chromosome of a given length is between $O(n^2)$ and $O(n^3)$, where n is the chromosome length.

The GC problem is closely connected with construction of independent or internally steady subsets (IS), and also determination of cliques in a graph. If any two nodes of a subset X' of a graph G = (X, U), $X' \subseteq X$, are not adjacent, X' is referred to as internally steady. A subset $\Psi_i \subseteq X$ of a graph G = (X, U) is referred to as a maximal internally steady or independent, if adding to it any node $x_i \subseteq X$ makes it not internally steady. A subset Ψ_i is independent, if

$$\forall x_i \in \Psi_i (\Gamma x_i \cap \Psi_i = \varnothing). \tag{2.46}$$

IS differ as to the numbers of elements included in them. It is possible to determine a family of all IS, denoted $\Psi_{i=} \{\Psi_1, \Psi_2, \ldots, \Psi_s\}$, or its part, for any graph G [41, 66]. The GS used for GC are also effectively used in construction of IS families. For example, we shall construct an IS family for the graph G of Fig. 2.39. Regularization of nodes according to the increasing local degrees is initially done. We shall illustrate the functioning of the modified GCO on the example.

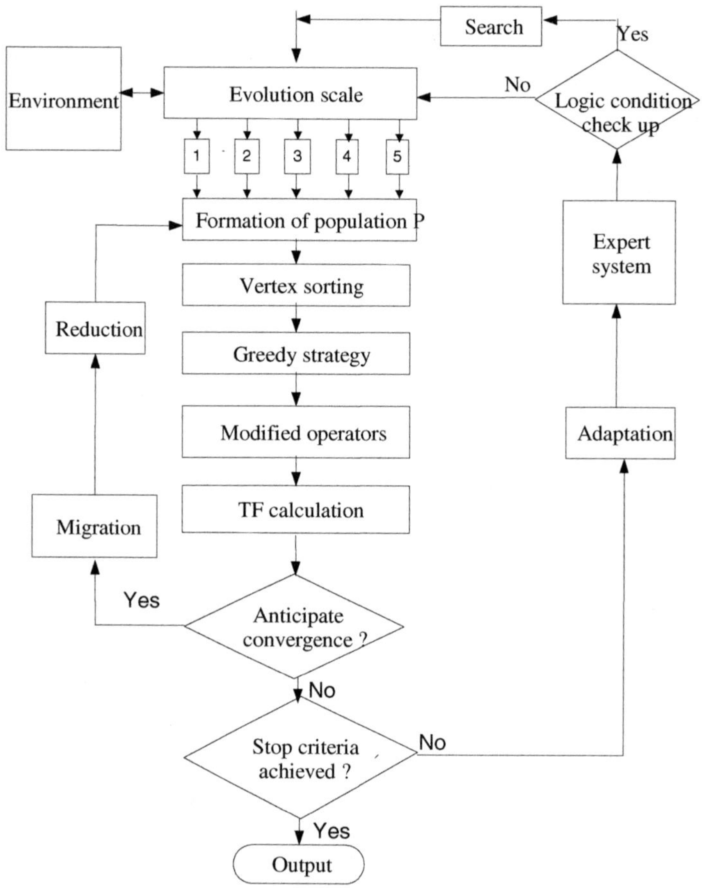

Fig. 2.38. Structure of hybrid GA

Let for the graph G of Fig. 2.39 the population P = {P_1, P_2, P_3, P_4, P_5} be given:

P_1: 1 2 3 4 5 6 7 8 9,

P_2: 9 8 7 6 5 4 3 2 1,

P_3: 6 1 5 4 7 9 3 8 2,

P_4: 4 9 3 7 8 1 7 6 2,

P_5: 5 6 7 8 9 2 3 4 1.

The algorithm starts by selecting in the ordered chromosome P_5 the first element, 5. Then, we select node 6, as next to node 5 in chromosome P_5 and not adjacent to it. In a similar manner we select further seven nodes in this chromosome. Afterwards, having analyzed population P, we select node 9 in chromosome P_3. Further analysis of population P shows that the first IS constructed is $\Psi_1 = \{5, 6, 7, 9\}$. There are a lot of strategies of constructing IS. One of them consists in construction of independent BB composed of two, three, four etc. nodes, to then associate these BB into IS. Another strategy (width-oriented) provides for the identification of all the IS, where for the first examined node (in our case node 5) the analysis is carried out of other nodes potentially entering the IS. The third strategy (depth-oriented) consists in consecutive analysis of all nodes from an ordered list (chromosome P_5) before determining the IS family. The fourth strategy is a combination of the first three ones. By applying the fourth strategy to graph G of Fig. 2.39 we obtain the set of IS $\Psi_2 = \{6,7,8\}$, $\Psi_3 = \{7,8,2\}$, $\Psi_4 = \{8,1,3\}$, $\Psi_5 = \{9,2,4,5\}$, $\Psi_6 = \{1,2,5\}$, $\Psi_7 = \{2,5,7,9\}$ $\Psi_8 = \{3,5,7,9\}$, $\Psi_9 = \{4,5,6,9\}$. It is obvious that this set is not complete. In order to obtain the complete IS family it is necessary to continue realization of the strategy proposed. The technique is effective for GC. For example, let Y_5 correspond to the first color, $X_1 = \{9,2,4,5\}$, Y_2 to the second color, $X_2 = \{6,7,8\}$, and $Y_{4\backslash 8}$ to the third color, $X_3 = \{1,3\}$. So, graph G of Fig. 2.39 is colored by three colors. The ART of construction of an IS family is between O (n^2) and O (n^4).

Now, let us note that the problem of constructing cliques in a graph is opposite to construction of IS [41, 66, 84]. The subset of the maximum number of mutually

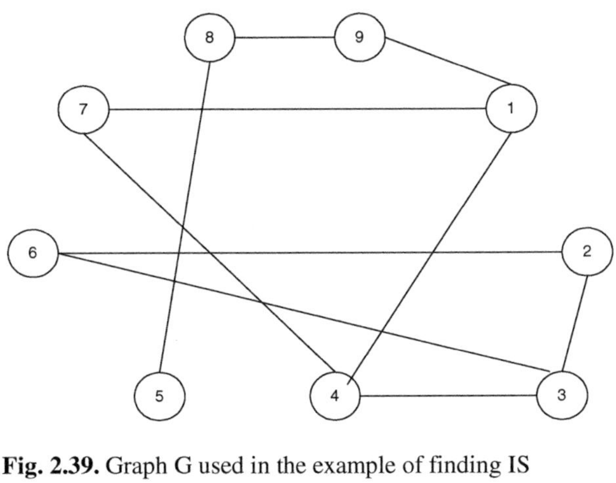

Fig. 2.39. Graph G used in the example of finding IS

adjacent nodes in a graph is referred to as clique. In a graph, it is in general possible to determine a family of cliques: $K = \{K_1, K_2 \ldots K_z\}$. It is obvious that the described strategy and GA can be applied for determining the clique family. Let us show on the example of graph G from Fig. 2.40 a simplified scheme of GA for determination of cliques.

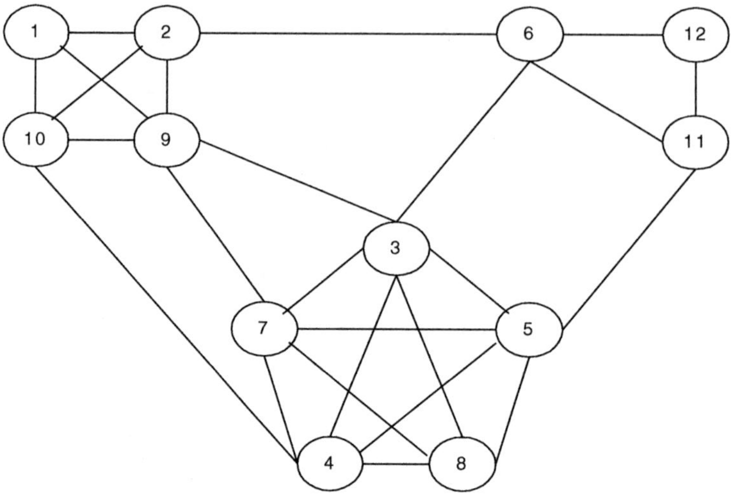

Fig. 2.40. Graph G used in the example of cliques determination

Let for the graph G of Fig. 2.40 a population $P = \{P_1, P_2, P_3, P_4, P_5\}$ be given:

P_1: 1 2 3 4 5 6 7 8 9 10 11 12,

P_2: 12 11 10 9 8 7 6 5 4 3 2 1,

P_3: 2 4 6 8 10 12 1 3 5 7 9 11,

P_4: 1 5 9 2 6 10 3 7 8 4 11 12,

P_5: 1 10 12 6 2 9 11 4 8 7 3 5.

Let us apply the following search scheme, based on GA and GS:

1. 1.
2. 1, 2 – a complete subset of two nodes is formed.
3. 1, 2, 3♦ (impasse).
4. 1, 3♦.
5. 1, 5♦.
6. 1, 10, 11♦.
7. 1, 10, 9, 11♦.
8. 1, 10, 9, 8♦.

9. 1, 10, 9, 2, 3♦.
10. 1, 10, 9, 2, 4♦.
11. 1, 10, 9, 2, 6♦.

The first clique, K_1 = {1, 10, 9, 2}, was constructed. The elements corresponding to nodes K_1 are removed from all chromosomes of the population. We obtain a new population with chromosomes of a shorter length:

P_6: 3, 4, 5, 6, 7, 8, 11, 12,

P_7: 12, 11, 8, 7, 6, 5, 4, 3,

P_8: 4, 6, 8,12,3, 5, 7, 11,

P_9: 5, 6, 3, 7, 8, 4, 11, 12,

P_{10}:12, 6, 11, 4, 8, 7, 3, 5.

By continuing, we get:

1. 3.
2. 3, 4.
3. 3, 4, 5.
4. 3, 4, 5, 6♦.
5. 3, 4, 5, 12♦.
6. 3, 4, 5, 7, 8, 11♦.
7. 3, 4, 5, 7, 8, 6♦.
8. 3, 4, 5, 7, 8, 12♦.

The second clique, K_2 = {3, 4, 5, 7, 8}, is determined. By proceeding similarly, we construct the third clique, K_3 = {6, 11, 12}. We shall note that in 21 elementary steps of the modified greedy strategy GS the basic three cliques of the graph were obtained. ART is O(n) in the best case and O(n!) in the worst case. On the average, for real graphs, it is at $O(n^2)$-$O(n^3)$.

A large number of combinatorial-logical tasks and OT on graphs require establishing isomorphism or isomorphic mappings between the structures considered [56, 57, 100—104]. This problem, like all other problems considered above, is NP-complete. Therefore, various heuristics have been developed meant to obtain acceptable practical results.

The problem of graph isomorphism recognition (GIR) consists in the following. For a given pair of graphs $G_1 = (X_1, U_1)$ and $G_2 = (X_2, U_2)$ we want to establish, whether there exists a mutually unique mapping φ: $X_1 \rightarrow X_2$ such that $U = (x, y) \in U_1$ if and only if $(\varphi(x), \varphi(y)) \in U_2$ [56].

The polynomial algorithms are known for this problem for the following graph classes: graphs of limited degree; graphs with limited frequency rate of eigenvalues; k-divisible graphs; k-associative graphs, etc. [56].

It is known that the problem of isomorphic mapping of a graph is an NP-complete problem [66]. It has a lot of similarity with, and at the same time it essentially differs, in terms of complexity, from GIR. For example, for solving a problem of isomorphism of a sub-graph G_1 with the use of known algorithms for GIR it is necessary to develop a procedure of determining in graph G of a subset $X_1 \subset X$ equipotent with the set of nodes X_2 in graph G_2.

Such a procedure includes k1 actions, where $k_1 = \binom{n}{n_2}$, $n = |X|$, $n_2 = |X_2|$.

Hence, it is necessary to apply k_1 times the GIR algorithm. Therefore, to determine each sub-graph G_1 in graph G it is necessary to carry out k_2 actions, where $k_2 = \binom{m_1}{m_2}$ (m_1 is the number of edges in sub graph G_1, and m_2 - in graph G_2, $m_1 > m_2$). Hence, even the GIR algorithm is polynomial, it is impossible to solve the problem of isomorphic mapping with its use in polynomial time [66]. It can be solved in polynomial time if G_1 is a wood, and G_2 is a tree.

Establishment of isomorphism of homogeneous graphs, which have automorphic sub-graphs [41], represents the biggest computational burden. The methods of graph partitioning referring to various levels are used for solving such problems. Thus, the running time of the algorithm is reduced by O (n!) to O (k!), where n is the number of elements in the graph, and k is the number of elements in the biggest automorphic sub-graph, i.e. in a sub-graph, for which the choice of nodes for establishing conformity has no significance.

The basic idea of such algorithms consists in the following. In the graphs analyzed with respect to isomorphism, selection of two isomorphic nodes is assumed. With regard to these two nodes, all the remaining nodes are divided into two classes. The first class includes nodes, adjacent to those assumed to be isomorphic, and the second – the remaining ones.

For example, Fig. 2.41 shows two graphs, G and G'. Let G = (X, U), G' = (X', U'), |X| = |X'| = 6, |U| = |U'| = 9, local degrees of all nodes of the graphs are equal three. It is necessary to establish the isomorphism of graphs G and G'. In other words, it is necessary to establish, whether there exists a relation of equivalence:

$$\begin{aligned}&X \Leftrightarrow X',\ U \Leftrightarrow U' \text{ such, that,}\\ &\text{if an edge } (x_i, x_j) \Leftrightarrow \text{an edge } (x'_i, x'_j),\\ &\text{then } x_i \Leftrightarrow x'_i,\ x_j \Leftrightarrow x'_j,\ x_i, x_j, \in X,\ x'_i, x'_j, \in X'.\end{aligned} \tag{3.22}$$

Let us show on this example the use of the heuristics of partitioning for establishment of isomorphism of homogeneous graphs. Let us select one node in the graph G and one in the graph G' (nodes numbered 1 in both cases). With respect to them the following partitioning can be done:

$\{1\}$ $\{2, 4, 6\}_{1+}$ $\{3, 5\}_{1-}$ (G),

$\{1'\}$ $\{4', 5', 6'\}_{1'+}$ $\{2', 3'\}_{1'-}$ (G').

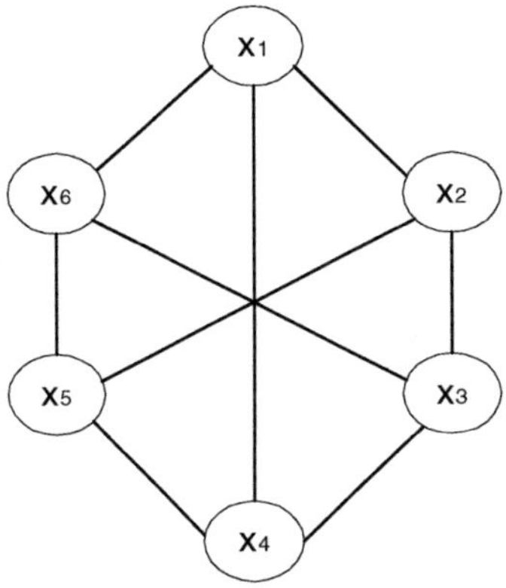

Fig. 2.41. a. Graph G

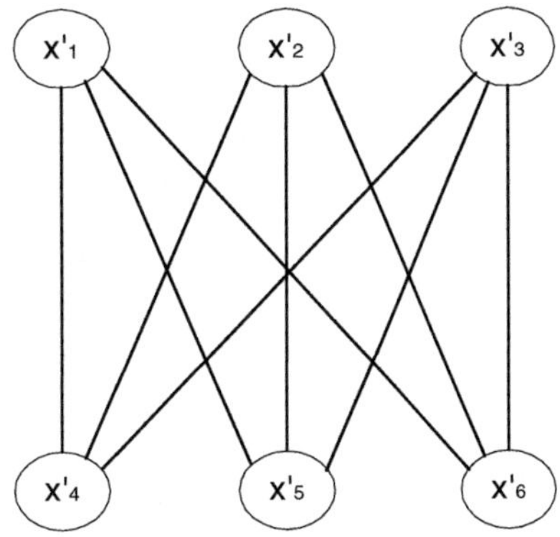

Fig. 2.41. b. Graph G'

In the above partitioning, the nodes x_2, x_4, x_6 are adjacent to node x_1, and nodes, x_3, x_5 are not adjacent to node x_1 in graph G. A similar partitioning is done for graph G'. (We assumed that nodes x_1 and x'_1 are isomorphic (P-isomorphic).) Then, subsets of smaller capacity are selected and inside them the contiguity of nodes is checked. In our example, nodes x_3, x_5 and x'_2, x'_3 are not adjacent to the initially chosen ones. Therefore, the process of partitioning continues:

$\{1\} \quad \{3\}_{1-} \quad \{\{2, 4, 6\}_{3+}\}_{1+} \quad \{\{5\}_{3-}\}_{1-},$

$\{1'\} \quad \{2'\}_{1'-} \quad \{\{4', 5', 6'\}_{2'+}\}_{1'+} \{\{3'\}_{2'-}\}_{1'-}.$

The result of the process is altogether that graph G is isomorphic to graph G'. The correspondence of nodes is as follows:

$$t = \begin{Bmatrix} 1 & 2 & 4 & 6 & 3 & 5 \\ 1' & 4' & 5' & 6' & 2' & 3' \end{Bmatrix}.$$

In the example here considered, the subsets of nodes $\{x_2, x_4, x_6\}$ and $\{x'_4, x'_5, x'_6\}$ are automorphic, i.e. each node in one sub-graph can be isomorphic to any node of the other sub-graph. In this example it was shown that instead of the ART for GIR equal O(6!) we get ART of O(3!). In the algorithms of this type it is necessary to carry out search among the elements of automorphic sub-graphs. It follows from the example that such a search is carried out consistently for all nodes, except for the last one in the sub graph.

We propose that in the here considered method for GIR, after partitioning of the graph to carry out all the modified GO for the subsets with the biggest cardinality. This allows for analyzing in parallel all subsets of automorphic nodes and improves the time of finding the results.

For example, in Fig. 2.42, a and b, non-trivial, non-uniform ("hard" [84]) graphs G and G' are shown. In these graphs |X| = |X'| = 7, |U| = |U'| = 8, $\rho_1 = 1$, $\rho_2 = 3$, $\rho_3 = 3$, $\rho_4 = 3$, $\rho_5 = 3$, $\rho_6 = 2$, $\rho_7 = 1$; $\rho_a = 1$, $\rho_b = 3$, $\rho_c = 3$, $\rho_d = 3$, $\rho_e = 3$, $\rho_f = 2$, $\rho_g = 1$. Node 1 can be P-isomorphic to node *a* or *g*. In the first case this leads to an impasse, while in the second case – to a successful move. We assume that node 1 is P-isomorphic to node g. Then, from the connectivity of the graph and the partitioning heuristics it follows that node 7 is P-isomorphic to node a. Then, it follows that node 6 is P-isomorphic to node f. Further, it is necessary to establish whether there is correspondence between the subsets of automorphic nodes {2, 3, 4, 5} and {b, c, d, e}. In this case it is necessary, as noted above, to carry out, in general, a complete search for these subsets of nodes. Here, the respective ART is 4! Upon the performance of these operations we get that node 2 is P-isomorphic to node b, node 3 is P-isomorphic to node d, 4 to c, and 5 to e. Hence, graphs G and G' of Fig. 2.42 a, b, are isomorphic, and the correspondence of nodes is:

$$t = \begin{Bmatrix} 1 & 2 & 3 & 4 & 5 & 6 & 7 \\ a & b & d & c & e & f & g \end{Bmatrix}.$$

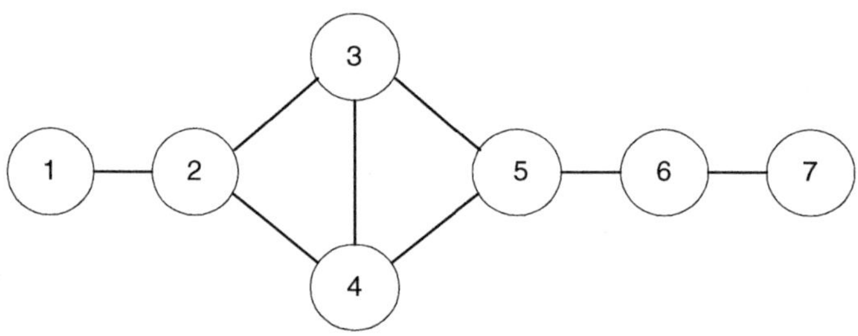

Fig. 2.42. a. Non trivial graph G

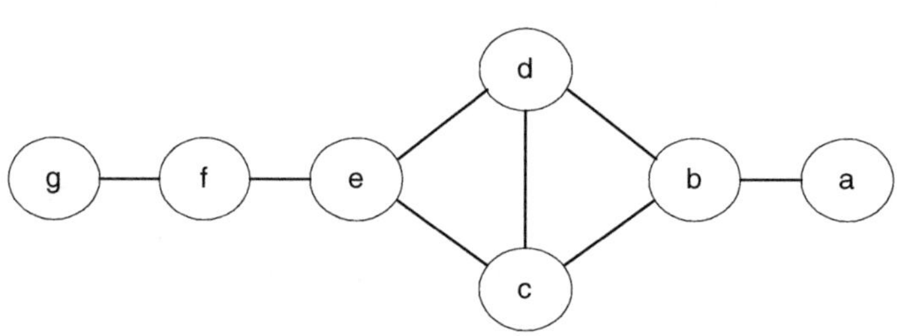

Fig. 2.42. b. Non trivial graph G'

Apparently, the presence of nodes with various local degrees simplifies the process of GIR. It allows for selection of initial conditions for the partitioning heuristics. In the example of Fig. 2.42, a, b, there are only two ways of partitioning to choose from, while in case of Fig. 2.41, a, b, there are six possible ways of partitioning. In homogeneous graphs, all degrees of nodes are identical; therefore, initial conditions are established in a random manner. So, for establishment of isomorphism in homogeneous graphs ART equals in the best case O(n), and in the worst case - O (n!). If graphs are not isomorphic, it is impossible to establish the result at one iteration. It is necessary to carry out the isomorphism-oriented comparison of one randomly chosen node of one graph with all nodes of another graph.

Consider, for example, two graphs of Fig. 2.43, a, b, G and G'. Here, G = (X, U), G' = (X', U'), $|X| = |X'| = 6$, $|U| = |U'| = 9$, and local degrees of all nodes the graph are equal three.

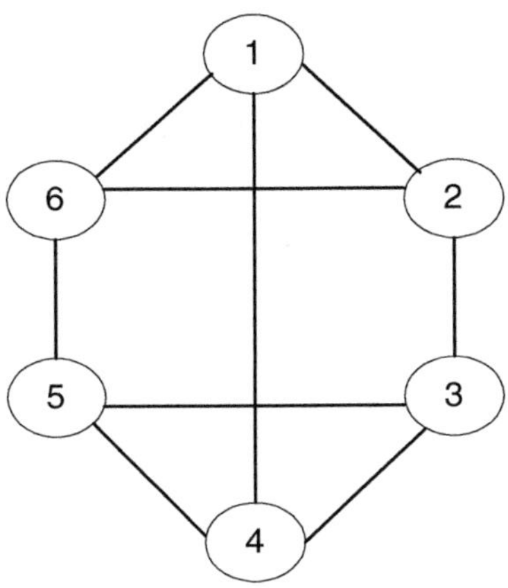

Fig. 2.43. a. Graph G

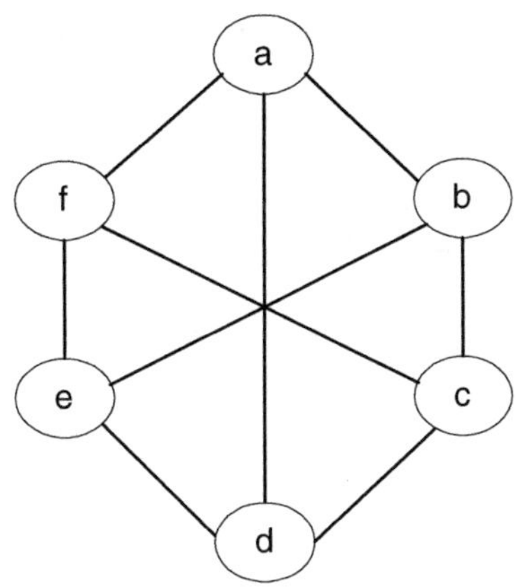

Fig. 2.43. b. Graph G '

To solve GIR, we apply the partitioning heuristics. We assume that node 1 of graph G is P-isomorphic to node *a* of graph G'. Then we obtain:

$\{1\}$ $\{2, 4, 6\}_{1+}$ $\{3, 5\}_{1-}$ (G),

$\{a\}$ $\{b, d, f\}_{1'+}$ $\{e, c\}_{1'-}$ (G').

For further analysis we choose the corresponding smallest subsets, {3, 5} and {e, c}. Nodes (3, 5) in graph G (Fig. 2.43a) are adjacent, while edge (c, e) in graph G' does not exist. Hence, node 1 cannot be isomorphic to node a. In this connection it is necessary to carry out similar operations to check the isomorphism of node 1 with other nodes of graph G '. Then, we obtain:

$\{1\}$ $\{2, 4, 6\}_{1+}$ $\{3, 5\}_{1-}$ (G),

$\{b\}$ $\{a, c, e\}_{1'+}$ $\{d, f\}_{1'-}$ (G').

Nodes (3, 5) in graph G (Fig. 2.43a) are adjacent, while edge (d, f) does not exist in graph G'. Hence, node 1 cannot be isomorphic to node b. By continuing similar constructions, we get that node 1 cannot be isomorphic to nodes c, d, e. Finally, suppose that node 1 of graph G is P-isomorphic to node f of graph G '. Then we obtain:

{1} $\{2, 4, 6\}_{1+}$ $\{3, 5\}_{1-}$ (G),

{f} $\{a, c, e\}_{1'+}$ $\{b, d\}_{1'-}$ (G').

Nodes (3, 5) in graph G (Fig. 2.43a) are adjacent, while edge (b, d) does not exist in graph G '. Hence, node 1 cannot be isomorphic to node b. All nodes of graph G' are analyzed for isomorphism with node 1 of graph G. In all cases the answer is negative. So, graph G is not isomorphic to graph G'.

Let us consider the example of GIR for graphs G and G' of Fig. 2.44, a, b. Here, G = (X,U), G' = (X',U'), |X| = |X'| = 10, |U| = |U'| = 15, local degrees of all nodes of both graphs are equal to three. We assume that node 1 of graph G is P-isomorphic to node *a* of graph G'. Then we obtain:

{1} $\{2, 5, 6\}_{1+}$ $\{3, 4, 7, 8, 9, 10\}_{1-}$ (G),

{a} $\{b, e, f\}_{1'+}$ $\{c, d, h, g, k, i\}_{1'-}$ (G').

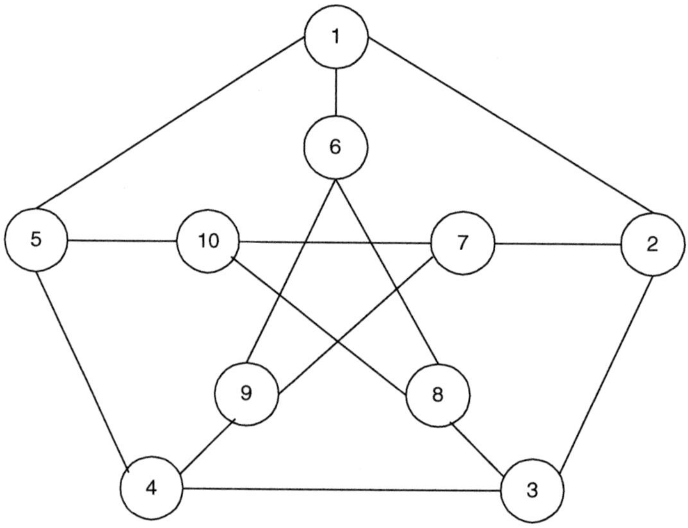

Fig. 2.44. a. Graph G

For further analysis, corresponding smallest subsets, {2, 5, 6} and {b, e, f} are selected. We assume that node 2 is P-isomorphic to node b. We get:

{1} $\{2\}_{1+}$ $[\{5, 6\}_{1+}]_{1-}$ $[\{7, 3\}_{1+}, \{4, 8, 9, 10\}_{1-}]_{1-}$ (G),

{a} $\{b\}_{1'+}$ $[\{e, f\}_{1'+}]_{1'-}$ $[\{g, c\}_{1'+}, \{d, h, k, i\}_{1'-}]_{1'-}$ (G').

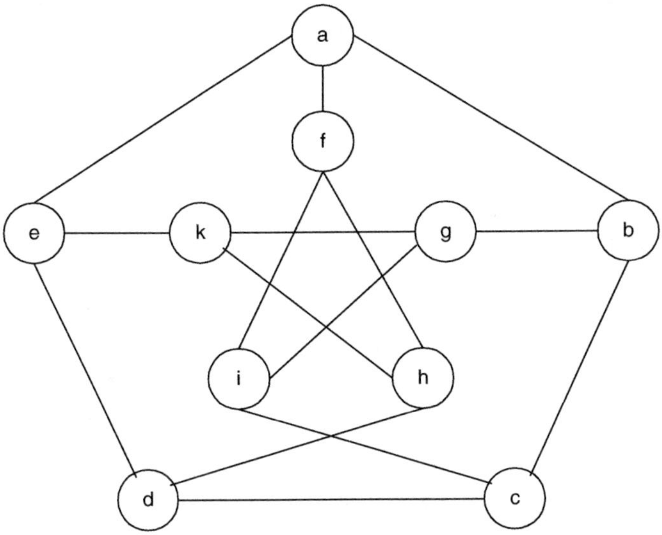

Fig. 2.44. b. Graph G'

Apparently, P-isomorphism holds. Continuing the procedure, we obtain:

$\{1\} \quad \{2\}_{1+} \quad [\{5, 6\}_{1+}]_{1-} \quad [[\{7\}_{1+}, \{3\}_{1+}]_{1+},[\{9, 10\}_{1-},\{4, 8\}_{1-}]_{1-}]_{1-}$ (G),

$\{a\} \quad \{b\}_{1'+} \quad [\{e, f\}_{1'+}]_{1'-} \quad [[\{g\}_{1'+}, \{c\}_{1'+}]_{1'+},[\{I, k\}_{1'-},\{h, d\}_{1'-}]_{1'-}]_{1'-}$ (G').

As nodes (h, d) in graph G' (Fig. 2.44b) are adjacent, and the edge (4, 8) in graph G (Fig. 2.44a) does not exist, node 1 cannot be isomorphic to node a, and node 2 to b, node 7 to g, node 3 to c. By so proceeding, we conclude that graph G is not isomorphic to graph G'.

The triune approach, based on micro-, macro- and meta-evolution is proposed for finding the isomorphic correspondence of graphs, if it exists. At the stage of micro-evolution the heuristics of partitioning allows for obtaining the BB. Redistribution of the genetic material on the basis of GO is carried out inside each BB. Owing to this, the size of chromosomes in a population decreases and the time of realization of the basic algorithm is reduced. For example, in chromosome P_1: 2, 3, 4, 5, which is a BB for graph G of Fig. 2.42a, we define mutation point between genes 3 and 4. Having executed the SMO, we obtain a descendant chromosome P_2: 2, 4, 3, 5. This determines the correspondence 2-b, 4-c, 3-d and 5-e.

At the level of macro-evolution it is effective to use FS, so that every BB is represented at a higher level as a new node of the graph. Hence, the size of chromosomes decreases as well and there appears an opportunity to increase the number of generations of GA in order to get the optimum result. It is especially important in the analysis of homogeneous non-isomorphic graphs with identical number of nodes and edges.

At the level of meta-evolution there is migration of chromosomes from one population to another and application of various modified GO to these populations.

Let us present the block diagram of GS for solving combinatorial logical search problems on graphs (Fig. 2.45), based on information feedback and the concept of incorporated evolution with genetic search management systems (GSMS).

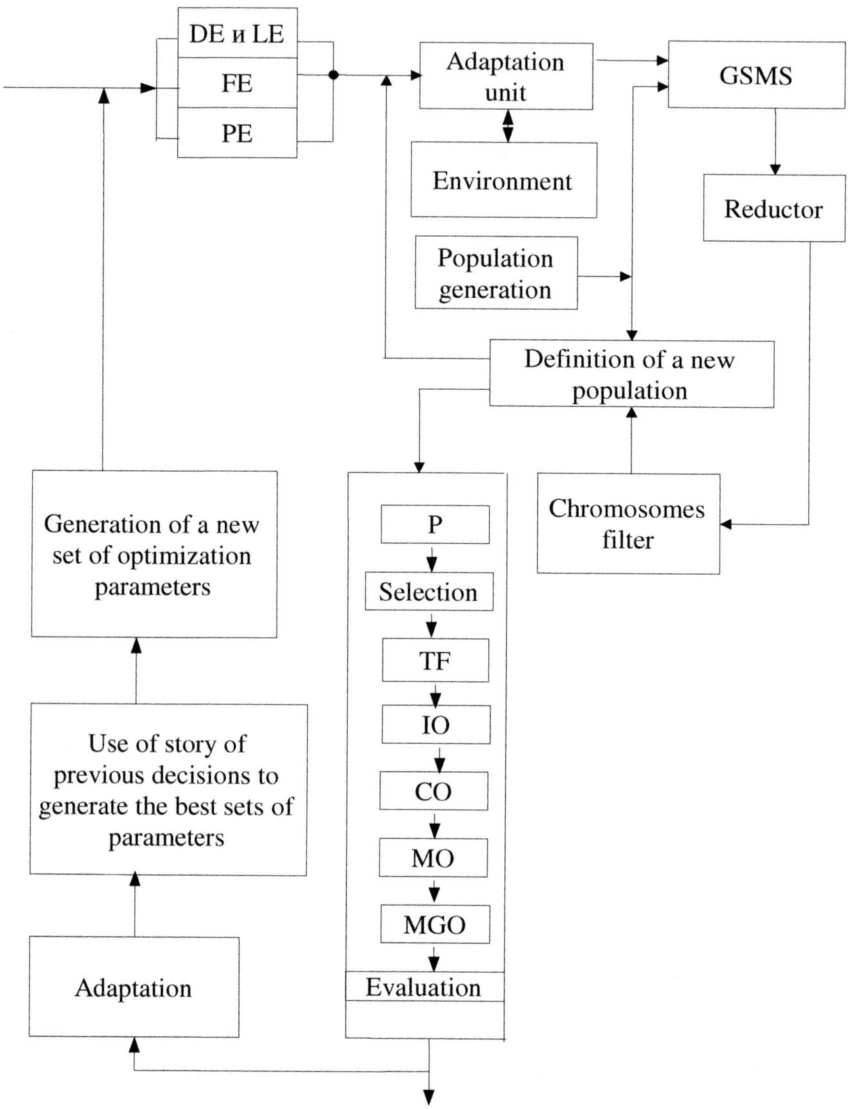

Fig. 2.45. Block diagram of managing the genetic search

After realization of GA, according to the scheme of Fig.2.45, adaptation unit and reductor, upon the interaction with the environment, implements synergetic principles, and the chromosome filter supports homeostasis. Thus, the best chromosomes are used in population transformations and in leaving the local optima. The reducer block shrinks the population, eliminating chromosomes with TF values lower than the average. The blocks of the adaptation unit, reductor and the chromosome filter allow for increasing the efficiency of realization of evolution and for speeding the GIR. It is necessary to note that in large graphs with non trivial automorphism (K>100) the process of establishment of isomorphism becomes much more complex, but the use of such search schemes reduces the order of running time.

In conclusion of this section we shall note that algorithms of graph coloring have been described, based on consecutive and greedy heuristics. Various ways of minimizing computational effort have been considered. Different modified GO are used for this purpose. Block diagram of the hybrid GA for graph coloring was shown. ART of algorithms in this class is contained between $O(n^2)$ and $O(n^3)$. The IS and clique oriented strategy of graph analysis has been considered, based on evolutionary approaches. In order to reduce the time of search in the establishment of isomorphism of homogeneous graphs the concept of joint partitioning and genetic search inside building blocks has been forwarded.

2.4 Construction and Analysis of the Evolutionary Algorithm for Finding Graph Pair Combinations

Use of evolutionary models leads in many respects to high quality of solutions to engineering problems. The analogy of evolutionary development of natural and artificial systems allows for the development of methods of evolutionary modeling, genetically distributed artificial intelligence and artificial life [29, 105]. Application of graph and hypergraph models enables solving engineering and practical problems more efficiently. One of the major combinatorial problems is finding the biggest pair combination. This problem is NP-complete [64, 66]. Hence, it is important to develop adequate heuristic algorithms. We propose here a new method of finding the biggest pair combination in two-part graphs, based on the combined genetic algorithms [37, 106]. This method, contrary to the known ones, allows for obtaining of a set of quasi-optimum solutions in sensible time.

The here considered process of evolution is founded on the analysis of initial population of alternative solutions and applies various kinds of evolutionary, genetic and combined algorithms. Genetic algorithms (GA) begin their work with creation of the initial set of alternative solutions. Then, these "parental" solutions generate "descendants" with the best properties through random, directed or combined transformations. Quality of each alternative solution is estimated and selection is performed. In all known models the principle of "survival of the strongest" or its modification is used, i.e. the least adapted solutions are eliminated, and the best ones pass to the following generation. The process is repeated until sets of optimum or quasi-optimum solutions are obtained [29, 37, 105, 106].

In solving optimization problems (OP) on graphs, GA offer many advantages. One of them is adaptation to changing environment. In GA a population is a source of knowledge, which can be analyzed, supplemented and altered with reference to changing conditions. For this purpose it is not necessary to carry out exhaustive search. Another advantage of GA is the ability to generate fast sufficiently good alternative solutions.

We propose a combined GA consisting of three blocks. The first block is the so-called preprocessor. It creates one or a set of initial populations, based on various methods of local search [37, 106].

Second block consists of four stages:

- choice of representation of the solution;
- development of operators of random, directed and combined changes;
- definition of conditions for the survival of a solution;
- recombination.

The third block is the so-called postprocessor. It implements the principles of evolutionary adaptation to the environment (the decision maker) and self-organization. The respective search scheme is shown in Fig. 2.46.

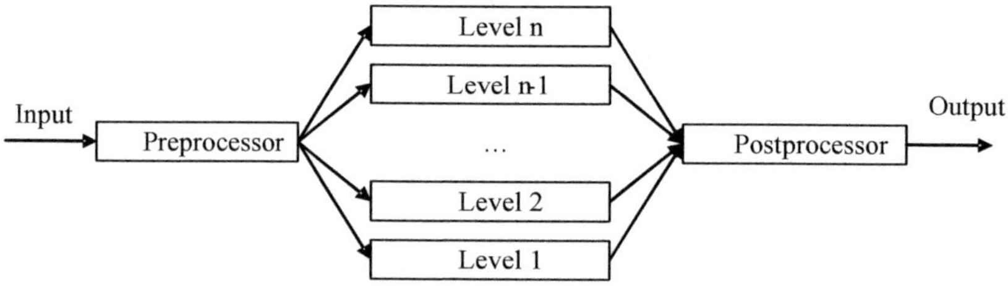

Fig. 2.46. Genetic search architecture

This is a horizontally organized architecture of genetic search (GS). Its advantage is that all levels are connected with the environmental adaptation level and can communicate with it. The disadvantage of horizontal architecture is complexity of coordination of work of separate levels. Here are the main principles of effective used of genetic search:

Principle of integrity. In genetic algorithms the value of the criterion function of alternative solutions does not reduce to the sum of values of the criterion function of partial solutions.

Principle of addition. When solving OP on graphs using GA it is necessary to use various models of evolution and various genetic operators.

Principle of discrepancy. With the growth of complexity of the analyzed problem the possibility of constructing exact model decreases. Here the theory of fuzzy graphs is applied.

Principle of conformity. The language of description of the initial problem should correspond to available information about it.

Principle of variety of ways of development. Implementation of a GA can be done in various alternative ways. There are many ways of evolution. The primary goal is to find the bifurcation point and to choose the way resulting in the optimum solution.

Principle of unity and contrast of order and chaos. "Chaos is not only destructive, but also constructive", i.e. chaos in the region of feasible solutions necessarily contains the order determining the required solution.

Principle of compatibility and divisibility. The process of evolution has a forward, pulsing or combined character. Therefore, a model of synthetic evolution should combine all these characters.

Principle of hierarchy. GA can be built from top down and from bottom upwards.

Principle of "Ockham's razor". It is undesirable to increase the complexity of the GA architecture needlessly.

Prigogine's principle of spontaneous emergence. GA allow for spontaneous generation of sets of alternative solutions, among which, with high probability, the optimum solution is contained.

Principle of homeostasis. GA are so designed that any produced alternative solution should be feasible. The GA operators should allow obtaining realistic solutions.

Let us write down the basic structure of the combined genetic algorithm for solving OP on graphs in view of the considerations presented:

1. Preprocessing
2. Creation of an initial population of solutions for OP on graphs.
3. Modeling of the population (finding TF for each chromosome) based on the principles listed.
4. Selection of the best chromosomes (alternative solutions) for implementation of genetic operators.
5. Implementation of the modified genetic operators when the criterion assumed is not satisfied.
6. Reduction (transformation) of the population to the type assumed.
7. Recombination of parents and descendants for a new generation.
8. Postprocessing.
9. Production of a new generation.
10. Termination of the algorithm.

2.4.1 Combined Genetic Algorithm for Finding Pair Combinations

Assume that G = (X,U) is a non-oriented graph. A pair combination (PC) is a subset of edges M⊆U that do not have common ends. And each edge $u_i \in U$ is adjacent to one edge from M. The maximum pair combination is an M that contains the biggest possible number of edges [64, 66]. It is known that the number of edges in PC of a graph G = (X,U) with |X| = n does not exceed [n/2], where [n/2] is the nearest integer not less than n/2..

Consider a new heuristics for finding pair combinations in two-part graphs G = (X1∪X2, U), X1∪X2 = X, X1 ∩ X2 = Ø.

Assume a two-part graph, shown in Fig. 2.47. It is possible to find, e.g., a pair combination M1 = {(1,6), (3,8)} in it. In this graph the biggest pair combination (LPC) is M2 = {(5,8) (3,7) (2,6)}, shown in bold lines in Fig. 2.48. We can build one more LPC in this graph: M2' = {(4,8) (3,7) (2,6)}. For finding LPC in two-part graphs we will use a special contiguity matrix R.

R =

	6	7	8	9
1	1			
2	1		1	
3		1	1	1
4			1	
5			1	

Rows of the matrix match vertices X1 and columns match vertices X2. The values of 1 or 0 denote the presence or absence of corresponding edges. This representation requires 4 times less memory cells for a matrix. This is especially important for analyzing very large graphs.

We propose the following heuristics. In matrix R a diagonal line with the biggest number of 1's is found. If there are more than one equivalent diagonal, any of them is selected. For example, in matrix R above the diagonal found is D = {(2,6) (3,7) (4,8)}. So, the LPC for the graph is established (Fig. 2.48).

Let us formulate the following hypothesis. The main diagonal of the matrix R can be completely filled with elements corresponding to the LPC. The total number of units corresponds to the total number of edges of the LPC. Each unit of the main diagonal determines an edge of the LPC. The proof follows from the way of constructing the special matrix, based on the two-part graph, because every edge from X1 on the main diagonal corresponds to one and only one vertex from X2. Note that before starting the algorithm it is necessary to order the vertices of the two-part graph in an increasing manner.

Let us describe the essential strategy of finding LPC in two-part graphs:

1. Define the subsets of vertices X1 and X2 of the two-part graph G.
2. Sort the vertices X1 and X2.
3. Build the special matrix R and define the main diagonal in it.

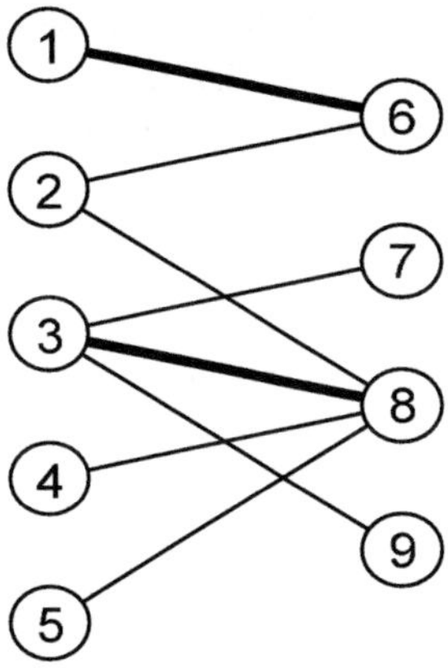

Fig. 2.47. Two-part graph G

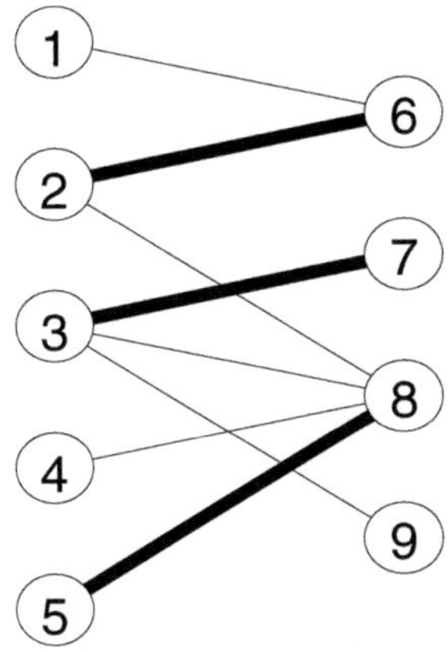

Fig. 2.48. The biggest PG (2,6), (3,7), (5,8)

4. If the main diagonal is all filled with 1's, the LPC is already determined and we go to step 7. Otherwise, we go to 5.
5. Define all the diagonals and select the one with the biggest number of 1's in matrix R. If there are more such diagonals, we choose any one of them.
6. Aggregate and transform with genetic algorithms.
7. Terminate the algorithm.

For example, let a two-part graph of Fig. 2.49 be given. We shall construct the special matrix R of this graph.

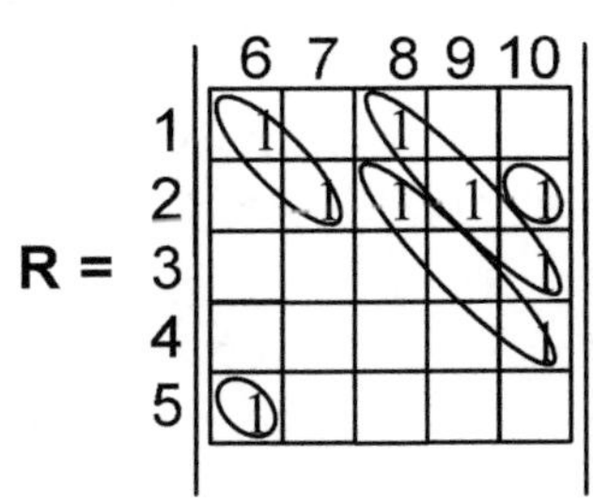

According to the algorithm we analyze the main diagonal D = {(1,6) (2,7)}. It is not completely filled with 1's. So, we look at other diagonals: D1 = {(1,8) (2,9) (3,10)}; D2 = {(2,10)}; D3 = {(2,8) (4,10)}; D4 = {(5,6)}.

We choose D1, as containing more elements than the other ones, to be the basis for finding the LPC. We perform aggregation: $D_1 \cap D_2 \neq \varnothing$, $D_1 \cap D_3 \neq \varnothing$, $D_1 \cap D \neq \varnothing$, $D_1 \cap D_4 \neq \varnothing$. So, we add element (5,6) from D_4 to D_1. This is the end of LPC building, M = {(1,8) (2,9) (3,10) (5,6)} and $|M_1| = 4$. This method can be used also for graphs that are not two-part graphs. For this, it is necessary to find the biggest two-part area in the graph, determine LPC for it and then, using aggregation and GO, determine LPC for an arbitrary graph [107, 108].

Take, for example, graph G of Fig. 2.50. For its two-part area the LPC is M = {(2,6) (3,7) (4,8)}. After aggregation we get M' = {(2,6) (3,7) (4,8) (5,9)}.

Note that to construct LPC one can use the greedy strategy. The example of such procedure is:

1. (1,6)
2. (1,6) (3,7)
3. (1,6) (3,7) (4,8)
4. (1,6) (3,7) (4,8)
5. (1,6) (3,7) (4,8) (5,6)
6. (1,6) (3,7) (4,8) (5,8)
7. (1,6) (3,7) (4,8) (5,9).

Thus, LPC is obtained, M = { (1,6) (3,7) (4,8) (5,9)}, |M| = 4.

For graphs with n<100 the set of all PC can be built and the biggest PC selected.

For example:

1. (1,7)
2. (1,7) (2,6)
3. (1,7) (2,6) (4,8)
4. (1,7) (2,6) (4,8) (5,9).

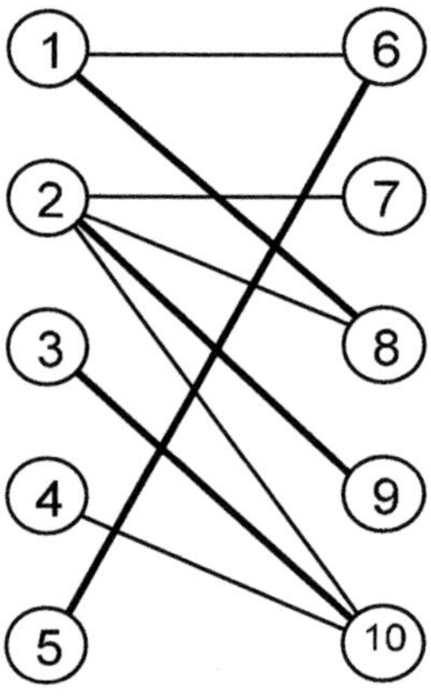

Fig. 2.49. Two-part graph G

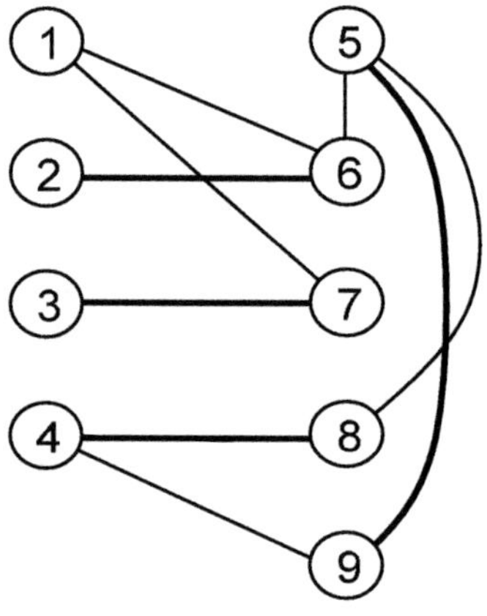

Fig. 2.50. Graph G

A new LPC is obtained, M1 = {(1,7) (2,6) (4,8) (5,9)}, |M| = 4.

Now, let us try out the greedy GO for the graph of Fig. 2.50 with chromosome length n/2 = 4½. So, L = 5. Generate randomly a population as a set of chromosomes consisting of different edges of the graph, P = {P1, P2, P3, P4}:

P1 = (1, 6) (2, 6) (3, 7) (4, 8) (5, 9),

P2 = (2, 6) (4, 9) (5, 8) (3, 7) (1, 7),

P3 = (7, 9) (5, 8) (1, 7) (2, 6) (5, 6),

P4 = (4,9) (5,9) (2,6) (4,8) (3,7).

Using segregation operator [37, 106] we choose a subset of edges (a building block) from P1: BB1 = {(4,8) (5,9)}, from P2: BB2 = {(1,7)}, from P3: BB3 {(2,6)}. At this step already construction of the LPC is complete: M = {(1,7) (2,6) (4,8) (5,9)}. Time complexity of the algorithm ranges from O(nm) to O(n!) in the worst case. On the average, for realistic graphs the range is $O(n^2)$ - $O(n^3)$, where n is the number of vertices and m is the number of edges of the graph.

The authors have developed a software system for solving optimization problems on graphs, using Borland C++ Builder and Visual C++. Debugging and

Table 2.3.

Size of graph, n	Number of generations (G)	Population size	Crossover probability (%)	Mutation probability (%)	TF (arbitrary unit)	Time (sec)
100	10	100	97	8	8438	4
100	20	100	97	8	8864	6
100	30	100	97	8	8887	10
100	40	100	97	8	8906	14
100	50	100	97	8	9017	16
100	60	100	97	8	8969	18
100	70	100	97	8	9033	20
100	80	100	97	8	9012	23
100	90	100	97	8	8978	27
100	100	100	97	8	9003	31

testing were done on IBM PC-class computers with Pentium-IV and AMD Athlon A (0)-1500 processors and 512 MB of RAM. The results of the experiments performed, showing the dependence of TF values attained and time to solution finding upon the number of generations, are shown in Table 2.3.

2.5 Conclusion

In solving optimization problems on graphs it is necessary to take into account the dependence of the quality of solutions obtained on the initial data. Therefore, for implementation of particular genetic algorithms the authors suggest to account for the influence of the environment, the principles of genetic search and knowledge of the problem solved.

New heuristics for finding the biggest pair combinations in two-part graphs have been developed. The use of proposed principles and search methods allows for obtaining a set of maximal pair combinations in times comparable with those of realization of the consecutive algorithms.

Chapter 3
Algorithms and Models of ECE Module Connections Routing

3.1 The Statement of the Routing Problem and a Classification of Algorithms

Connections routing; connections list; Prim, Steiner trees; flat laying; selection rule; wave, labyrinth, channel and flexible routing algorithms; routing discrete working field.

Connections routing constitutes usually the final stage of ECE engineering design. It consists in defining the lines connecting equipotential contact elements and components of the device designed.

The routing problem is among the most difficult ones in the automated ECE design. This is connected with several factors, particularly with the variety of connection methods. Each connection method has different optimization criteria and limitations. From the mathematical point of view, routing is the most complex selection problem among a great number of the potentially optimal solution variants.

Simultaneous optimization of all connections during routing, performed as exhaustive review of all variants, is practically impossible. So, the routing methods used perform, as a rule, local optimization, when routing is optimal only on a given step, when connections traced before are available.

The essence of the routing problem can be formulated as follows: for a given scheme of connections, build the system of the necessary conductors on a plane (circuit board, chip, etc.) to realize the scheme of electric connections, considering predefined limitations. Main limitations refer to conductor widths and minimal distance between them.

The source information for solving the connections routing problem is usually constituted by the list of circuits, parameters of construction elements and commutation field and element position data. Routing criteria can include percentage share of realized connections, overall length of conductors, number of layers, number of conductor intersections, uniformity of conductor distribution, minimal routing area, etc. These criteria are often exclusive, and so evaluation of routing quality is performed by the dominating criterion, with other criteria treated as constraints. Also, additive (or multiplicative) evaluation functions are used like:

$$F := \sum_{p} \lambda_i \cdot f_i ,$$

where p is a number of partial criterion, F is an additive criterion; λ_i (i = 1,2, …, p) is a partial criterion; f_i is the weight of the partial criterion ($f_i + f_2 + \ldots + f_p = 1$).

V.M. Kureychik et al.: Genetic Algorithms for Applied CAD Problems, SCI 212, pp. 115–191.
springerlink.com

The routing problem has always the *topological* and *metric aspects*. The topological aspect is connected with determination of admissible space of positions of separate connection fragments, considering also the limitations related to layers, intersections, etc. The metric aspect takes into account the structural dimensions of elements, connections and commutation field, and metric limitations to routing.

The routing problem can otherwise be conceived as four successive steps:

- definition of the list of connections;
- allocation of connections within layers;
- definition of the order of connections;
- routing of conductors.

In the first step an exact list is established, showing the connections (circuits or fragments of circuits) that should be traced. In the second step it is attempted to decide exactly where each connection can be placed. In the third step the order of connections for each layer is defined; it is shown when each conductor will be placed in this layer. In the fourth step the answer is given as to how each connection should be built.

Let us look through these steps. The source information for the first step is the list of connections of each element contacts with other element contacts. Single connections of contacts are written directly into the connections list (or a connections matrix). A difficulty arises when it is necessary to connect one contact with several contacts. To solve this problem algorithms are used for constructing shortest covering trees of *Prim* and *Steiner*.

In the second step the decisions are made as to what conductors should be traced.

In defining the list it is necessary to solve the question of assignment of element contacts to connector contacts. In practice, it is often attempted to exclude intersections when designing straight lines between element contacts and corresponding connector contacts. In routing, many elements have logically identical contacts. Therefore, for simplification of routing, conductors connecting such contacts are examined and, when necessary, replaced in order to eliminate intersections of straight lines.

After the definition of the list of routing connections, the problem of conductor allocation on layers is solved. It is reduced to the construction of the flat graph layout. The final variant of allocation in layers can yet be improved by the analysis of each conductor.

In the third stage examination of each layer is performed, to determine the exact moment, when each conductor can be traced. In other words: to define the order of conductor processing for each layer. There is a selection rule for two conductors: if a contact B appears in rectangle *A*, i.e. in a rectangle that has conductor contacts in the opposite corners, then conductor B should be traced first. Unfortunately, this rule cannot be applied in all situations. Let us recall the common *Eikers' heuristic rule* of the ordering in conductor routing. Conductors are traced in the order of their priority values. The priority value of a conductor *V* is equal to the number of contacts in rectangle *W* [84]. In practice, usually the short horizontal and vertical segments of conductors are traced first, followed by the almost horizontal and vertical conductors, c lose to them. Finally, long connections are traced that are often situated outside other conductors. Here A and B are contacts of circuits, W are rectangles built on these contacts.

It is common to consider the problem of allocation of connections in the layers as a special graph coloring problem. Afterwards, examination of each layer is performed so as to determine the order of connection processing for each layer. When all connections are allocated within layers and the order of their processing is determined, the problem of routing of all the circuit fragments is being solved in the fourth step.

The fourth step of routing is the essential and most complex one. It is decisive for the efficiency and quality of solution to the entire routing problem. Fig. 3.1 shows a classification of routing algorithms for this step.

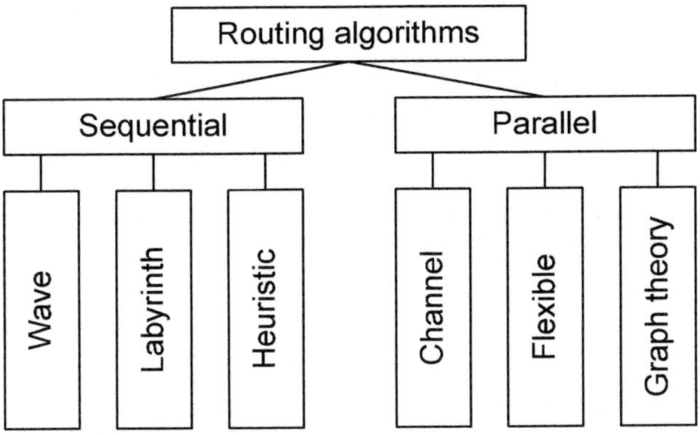

Fig. 3.1. A classification of routing algorithms

Wave algorithms are algorithms based on concepts of dynamic programming in the discrete working field (DWF). There exist now a great number of modifications of the wave algorithm, suggested by Li as early as in 1961. However, all these algorithms have one main drawback: high computational burden in terms of calculations and computer memory. Running time of the wave algorithm is $O(M N)$, where N is the number of DWF cells and M is number of points to be connected. Effective use of wave algorithms is possible only for DWF with cell numbers less than 10^5, and even then routing may take several hours of computer time. The number of untraced tracks is usually under 10%.

The essential concept of the *labyrinth algorithms* is to find a path between two DWF cells by sequential obstacle avoidance in the labyrinth, formed by the filled and free cells. Unlike wave algorithm, this one has direct character, so the search time decreases. The labyrinth algorithms usually build about 80% of connections.

Heuristic routing algorithms are usually effective only on the initial stages of routing, when there are a small number of obstacles on the plane. The main idea lies in generation of the longest free segments in the direction of the circuit with heuristic way of obstacle avoidance.

In *parallel routing algorithms* the admissible layout of tracks is determined first, then it is optimized with respect a selected criterion, and after that tracks are fixed on the commutation field.

Channel algorithms are treated nowadays as the having perspectives for the future. They are based on the (limited) possibility of regular fragmentation of the commutation field (CF) into channels and on routing vertical and horizontal segments inside channels. These algorithms are fast, but may have problems with solutions obtained. It is important to note that channel algorithms use a hierarchical approach to the routing problem. It means that the initial problem is decomposed into some sub-problems. For each hierarchical level there are corresponding routing methods, depending on the character of the level.

The first step in *flexible routing algorithms* is to build models of topology for circuit tracks in discrete topological working field (DTWF) (without strict fixation). In the second step, geometric models are built to determine configuration and allocation of each track on CF with metric accuracy.

The *graph theoretic* methods use graph models of scheme elements and components, and on their basis the topological pattern of the designed device is synthesized. In the next step the problem of evaluation of the topological pattern is solved.

Let us review the CF models for used in the algorithms mentioned. The wave and labyrinth algorithms use the following DWF:

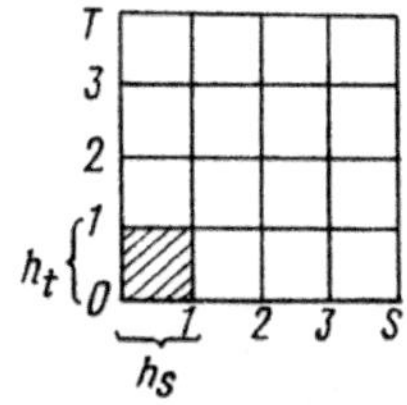

Fig. 3.2. CF model for the wave and labyrinth routing algorithms

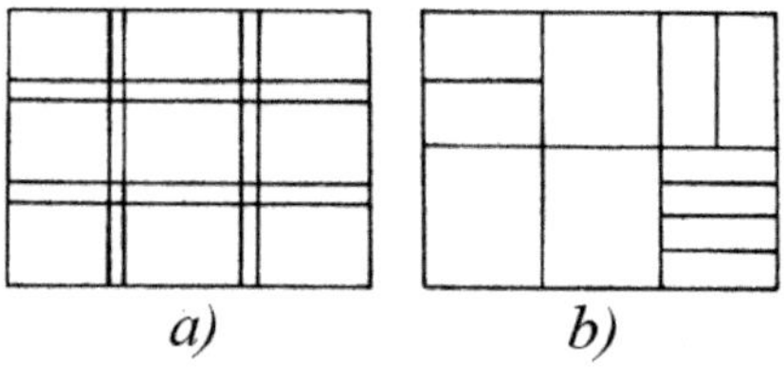

Fig. 3.3. CF model:

a — regular fragmentation;
b — irregular fragmentation

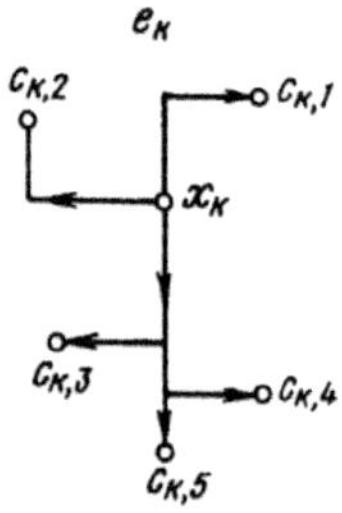

Fig. 3.4. Example of an oriented tree for circuit e_k

We have a rectangular coordinate system, designated SOT (Fig. 3.2), over a CF, with h_s and h_t units selected along the OS and OT axes. Then we draw lines, parallel to the axes, across the points αh_s and αh_t of the axes. Then, the CF plane can be divided into elementary rectangular cells sized $h_s \times h_t$, the dividing lines forming a coordinate grid with steps h_s for OS and h_t for OT. These cells form the DWF. The values of h_s and h_t are chosen on the basis of CF area, and the admissible density of element contacts and conductors.

The CF model for the channel and heuristic algorithms, as in the previous case, is divided into rectangular areas called channels, but the division rules are defined by allocation of element contacts. If elements are of the same type and are allocated on the plane regularly, division into channels will also be regular, otherwise it will be irregular (Fig. 3.3 *a*, *b*).

The length of the side of any channel parallel to the primary direction of layer tracks is called channel length, and the length of the other side - channel width. The number of tracks that can cross the side of the channel is called carrying capacity of the channel.

Regular division of CF into rectangles poses no problems. Irregular division is slightly more difficult. It is convenient to have as small number of rectangles in the division as possible, since this reduces routing time and the size of tables for storing data.

For the definition of track topology in flexible routing algorithms a DTWF is constructed. It can be presented as a graph $G_r = (X_r, U_r)$, where X_r corresponds to element contacts, typical points of board edge, external contacts, restricted areas; U_r - to lines that connect the elements $x \in X_r$.

Generally, all edges of any configuration can take discrete values. The edges not filled with tracks can shrink, and edges with big numbers of tracks can grow.

Now, let us provide a mathematical structure for the electric connections routing problem. The source data include the connections diagram (CD), the plane structure, and the results of element allocation. We assume that all tracks are built within one layer; CD defines the subsets of contact sets C_p ($p = 1, 2\ldots k$) with given coordinates; these sets do mutually not intersect. The contacts of each subset, $C_{p,I,}$ are connected by electric connection, i.e. a circuit. It is necessary to connect contacts of the circuit, avoiding intersection of different circuit fragments. The distance between fragments should not be less than a predefined value. The mathematical model of the structure is a regular coordinate grid (graph G_r) that corresponds to DWF. Note that DWF cell size is usually defined by the area of the routing plane, and the limitations to contact and connection density. Let DWF cells be square shaped and each cell $x_i \in X \setminus X_3$ carry just one connection; X_3 is set of DWF cells, in which tracks should not be built. Each DWF cell corresponds to a vertex of $G_r = (X_r, U_r)$. Then, any circuit l_k can be represented as a directed tree $T_k \subset G$ (a CD graph) with root in an arbitrarily selected vertex x_k (Fig. 3.4). The T_k tree can be viewed as a union of simple paths, $(x_k,c_{k,1})\cup(x_k,c_{k,2})\cup(x_k,c_{k,3})\cup(x_k,c_{k,4})\cup(x_k,c_{k,5})$ here. Each path $(x_k, c_{k,\ r})$ consists of edges $u_{i,j} = (x_i, x_j)$ corresponding to tracks passing from cell x_i to x_j.

The following Boolean variables are introduced:

$y_{k\ ,j}^{\ i\ \ r}$ = {1 if edge $u_{i,j}$ belongs to path $(x_k, c_{k,\ r})$, otherwise 0}.

The circuit length is expressed in number of cells covered, i.e. G_r graph vertices, and is defined as

$$L_k = \sum_{i\in X_i \setminus X_3} \sum_{j\in X_j \setminus X_3} y_{i,j}^{k} + 1 \quad , \qquad (3.1)$$

where $x_j \in X_j$, and X_j is the set of cells, neighboring with i and belonging to $X_i \setminus X_3$;

$$y_{i,j}^{k} = \bigcup_{r=1}^{n_k} y_{i,j}^{k,r}, \qquad (3.2)$$

n_k = |T|-1 is the number of tree vertices.

On the basis of the above, the routing criterion is formulated as:

$$L_k = \sum_{k=1}^{m} L_k = \sum_{k=1}^{m} \sum_{i \in X_i \setminus X_3} \sum_{j \in X_j \setminus X_3} y_{i,j}^{k} + m.$$

For each path (x_k, $c_{k,r}$), the following limitations are usually set:

only one edge that belongs to a path can come out from a vertex x_k;
no edges should come into each vertex x_k;
vertices $c_{k,r}$ have only one incoming edge from path (x_k, $c_{k,r}$);
vertices $c_{k,r}$ do not have outgoing edges;
there should be no path breaks at all vertices $i \in X \setminus X_3$.

For calculation of different tracks the constraint of non-crossing is introduced

$$\sum_{i \in x_j} \sum_{k=1} y_{i,j}^{k} \le 1. \qquad (3.3)$$

Minimization of criterion L, taking into account the limitations described, can be treated as an IPP (integer programming problem); its solution will define the routing of connections. Note that because of the cumbersome and lengthy solving procedure, in practical problems heuristic routing algorithms with minimization of the described criterion or its modifications are used.

3.2 Wire Connections Routing

Wire connection; modified Prim and Hunan algorithms; Steiner point

Depending on the nature of construction realization of connections, the following types are distinguished: wire connections routing, printed connections routing, and routing of inter-circuit connections inside highly-integrated chips (LSI). We consider here the questions of wire connections routing and one-sided, two-sided and multi-layer circuit board assembling.

Among all the construction and technological realizations of connections, wire assembling is the simplest of the point of view of solving practical routing problems. This is due to the fact that conductors are isolated from each other and there is no problem of limiting their crossings. Therefore, the main routing criterion for the wire connections routing is the overall length of connections.

For wire assembling the routing problem is reduced to construction of trees with minimal total wire lengths in graph vertices. Here, graph vertices simulate connecting contacts, and edges - interconnections. The only limitation applies to the maximum degree of vertices, as it is impossible technologically to connect more than a certain number of connections to one contact.

The *minimal tree construction problem* is defined as follows. Let $P = \{p_1, p_2 \dots p_n\}$ be the set of points on the plane that correspond to points (contacts) of inputs and outputs of an arbitrary circuit. Let us examine a complete graph $K_n = (X, U)$, with vertices $x_i \in X$ corresponding to circuits and edges $u_j \in U$ edges - to assigned values of weight $\mu(u_j)$, describing connections between pairs of contacts. The value of $\mu(u_j)$ is a linear combination of several connection characteristics $\mu(u_j) = \mu_1 d_1(u_j) + \dots + \mu_s d_s(u_j)$, where values of μ_i are weight coefficients, and d_s is a characteristic of μ_j connection.

Now, the source problem reduces to search in graph G of the tree that includes all vertices $x \in X$ and has minimal total weight of all edges, i.e. the shortest covering tree (SCT). Solution to this problem is presented in the works of Kruskal, Loberman, Weinberger and Prim.

Take value $\mu(u_j)$, assigned to graph edges, for the distance between graph vertices. The minimum distance from separate vertices to a fragment is taken as distance between a vertex and an isolated fragment. The Prim algorithm for building the shortest covering tree or circuit with n contacts, can be described in the following way:

For an arbitrary circuit contact, find the nearest one and build a connection.

At each successive step $i = 2, 3 \dots n$ select from the subset of as yet unselected outlets the one that is closer than others to the group of the already connected outlets, and then connect the outlet to this group using the shortest path.

Here is an example (Fig. 3.5). The problem is to connect the equipotential contacts, marked "1", in a circuit. For this example, building of SCT using Prim algorithm is shown step by step in Fig. 3.5 b-e.

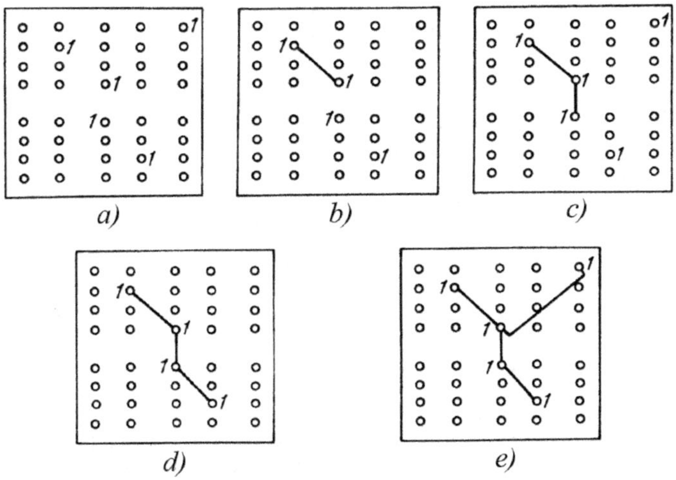

Fig. 3.5. Example of step-by-step construction of SCT

Construction of SCT using Prim algorithm has one quality: degree of any vertex does not exceed 6. As a rule, during development of assembling schemes of wired connections a limitation is applied on the maximum number of connections per one contact. For solution of such problems algorithms are often used based on the branch-and-bound method, though heuristic algorithms are preferred in the analysis of big numbers of circuits.

For example, Prim algorithm is examined here. Each isolated vertex is connected with the closest one that is not connected with this vertex by the intermediary of other vertices; each isolated fragment is connected by a shortest edge with the closest vertex that is not connected with it by other vertices.

In some cases, besides the limitation to the vertex degree, the start and end points of the circuit are defined. This is made, for example, during development of wiring diagrams for radio-frequency circuits, when it is necessary to connect signal source with some loads in a specified sequence. Sometimes the problem reduces to building the shortest path between two given contacts, passing through all other circuit contacts. This problem is related to the traveling salesman problem and, in terminology of graph theory, this is a problem of building the shortest Hamilton circuit between the predefined first and last vertices.

Here is an algorithm that produces approximate solutions to this problem. The algorithm is based on an (n-1)-step process of shortest edge selection in a complete graph K_n, testing of each edge for constraints, and composing a path, connecting definite points, of the edges selected and tested.

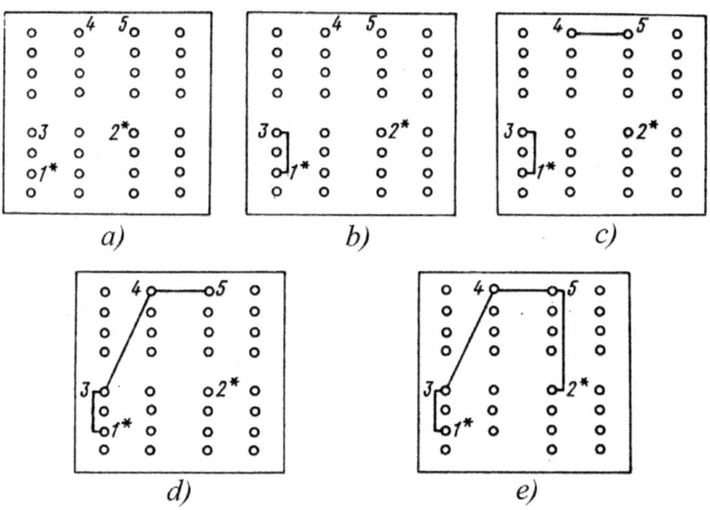

Fig. 3.6. Example of the step-by-step functioning of the algorithm

The example of Fig. 3.6 illustrates the algorithm. The locations of points to be connected on the board are shown in Fig. 3.6a. We have to compose a complete graph K_n, of the set of edges, ordered according to increasing length: (1* - 3),

(1* - 2), (2* - 3), (4 - 5), (3 - 4), (3 - 5), (1* - 4), (2* - 5), (2* - 4), (1* - 5). Points 1* and 2* are the first and last points of the path.

The successive edges $i = 1, 2\ldots n - 1$ are selected sequentially from this set considering the following conditions:

1. The edge does not connect the start and end points (1* and 2*).
2. When an edge is included in the path, the degrees of vertices do not exceed certain value λ (λ=1 for the first and last points and λ=2 for other points).
3. The added edge does not form a cycle with edges already in the path.
4. After the edge is included into path, the first and last points should stay unconnected, except for the n-1st edge.

Conditions 1-3 result directly from problem constraints. Condition 4 prevents the situations, when subsequent path forming becomes impossible, when all points except for the first and last have degree λ=2. The step-by-step process of path formation is shown in Fig. 3.6, *a—e*. The average difference of the length obtained, compared with optimal solution is at about 5%.

The problem of developing the SCT becomes more complicated, if during creation of the set of vertices (circuit contacts) one can use additional connection points. The respective problem can be formulated as: for definite surface points x_1, $x_2\ldots x_n$ an SCT with $n' \geq n$ vertices should be constructed. This problem is called *Steiner problem*. Usually, solution to this problem is sought in areas of rectangular structure. For $n \leq 5$ the solution to this problem is known, in general case there are conditions that Steiner trees (ST) and respective heuristic algorithms should meet. Additional points, added in construction of SCT, are called Steiner points (SP).

We list here the lemmas, formulated by Friedman and Menon, providing conditions for existence and construction of ST.

Lemma 3.1. For n vertices to be connected it is always possible to build an ST where each SP g_i is connected with at least three other points.

Lemma 3.2. For the set of vertices $\{x_1,x_2,x_3\}$ to be connected, the coordinates of SP g, minimizing $\sum_{i=1}^{3} d(g,n_i)$, are S_{avg} and T_{avg}, and the following holds:

$$\sum_{i=1}^{3} d(g,n) = 1/2P(x_1,x_2,x_3),$$

where S_{avg} and T_{avg} are the average values of S_i (i = 1, 2, 3) and T_j (j = 1, 2, 3); P(x_1, x_2, x_3) is the perimeter of the rectangle built over vertices x_1, x_2 and x_3; $d(g, n_i)$ is rectangular distance between g and n_i.

Fig. 3.7 shows an example illustrating the conditions of lemma 3.2 during construction of ST connecting vertices x_1, x_2 and x_3. The Steiner point g has coordinates $S_g = 3$, $T_g = 5$.

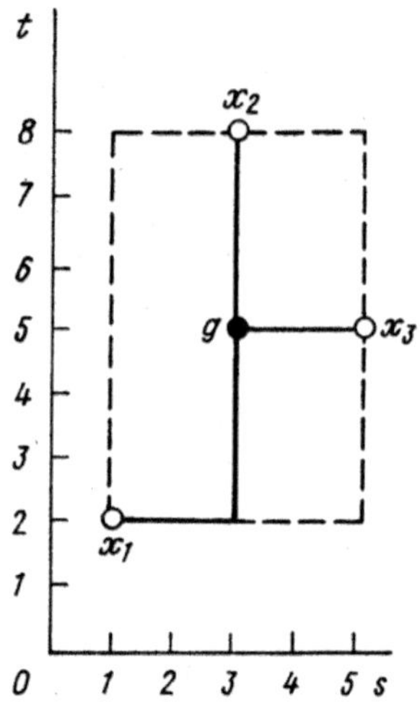

Fig. 3.7. Example of use of lemma 3.2

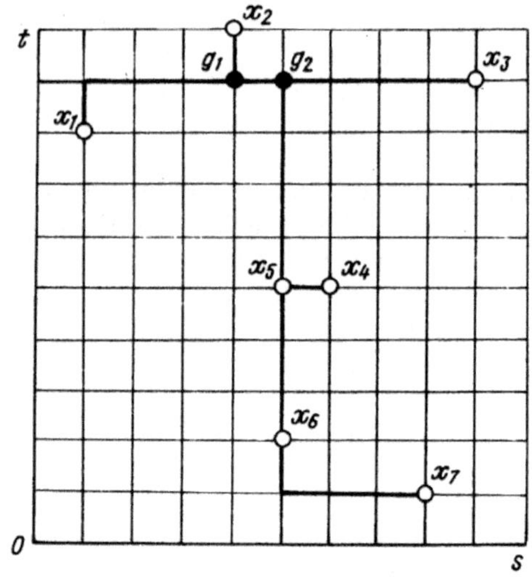

Fig. 3.8. Example of construction of ST

Lemma 3.3. Assume an ST consisting of x_i points on a plane. Let SP g be connected with only three vertices, x_1, x_2 and x_3. Then g is located inside the rectangle, built over points x_1, x_2 and x_3, and there is only one such g.

Corollary 3.1. For a definite set of vertices x_1, x_2, x_3 ST contains one SP g with coordinates (S_g, T_g) if it does not match with any of the vertices $\{x_1, x_2, x_3\}$ and the total length of ST edges is equal to $\frac{1}{2} P\,(x_1, x_2, x_3)$.

Lemma 3.4. Assume an ST over the set $X = \{x_1, x_2 \ldots x_n\}$ and SP of this tree forming a set $g = \{g_1, g_2 \ldots g_k\}$. If $g_1 \subset$ g and $g_i \subset I$, then g_1 contains an element $g_{1,k}$, connected with at least two vertices from I, otherwise g_1 is empty. Here, I is the set of points of coordinate grid line intersections.

On the basis of these lemmas a well-known theorem is formulated.

Theorem 3.1. Assume an ST over the set of vertices $X = \{x_1, x_2 \ldots x_n\}$ with the set of SP $g = \{g_1, g_2 \ldots g_k\}$. Then, for this set there exists another ST with the set of SP g', such that $(\forall g'_i \in g')\ (g'_i \in I)$.

From this theorem a method of SP detection, based on exhaustive search over all subsets of I points, can be derived. Obviously, this method can be applied only for building ST for small problems. This is the reason why heuristic procedures are developed to build the quasi-minimal ST.

Procedure 1

1. All vertices $x_i \in X$ are projected on the S (T) axis.
2. The distance between the extreme coordinates is divided by two, and from this point, $i \in I$ or $[i]$, if $i \notin I$, a perpendicular line is drawn (Steiner pole).

3. From each vertex $x_i \in X$ a line is drawn perpendicular to Steiner pole.
4. ST is constructed. Termination of the algorithm.

This algorithm is very simple, its running time is $O(n)$, but solution (in terms of total length of conductors) is far from perfect.

Procedure 2 (Hunan algorithm)

1. The set of vertices $x_i \in X$ is divided into classes S_0, S_1… S_i in increasing order of coordinates of S_i. All vertices of the same class should have the same coordinate S_i.
2. The class of vertices S_0 is analyzed, and all vertices of this set are connected by a perpendicular line with point $S_{\min}$ of axis S.
3. Set I' is formed, consisting of all points I that were previously connected with an arbitrary vertex x_i, including the vertices of ST construction.
4. Next, set S_{i+1} is selected, with the smallest value of S. Point $i_m \in I'$ and vertex $x_j \in S_{i+1}$ are defined, for which $d(i_m,x_j) \geq d(i_q,x'_j)$ and $\forall\ i_q \in I'$, $\forall x'_j \in S_{i+1}$. In other words, x_i is the closest to the examined ST fragment in the S_{i+1} class. Point i_m with coordinates (s_m, t_m) is connected to x_j vertex with coordinates (s_i, t_i) by a two-part broken line. The two parts are, respectively, parallel to axes S and T.
5. Operations similar to step 4 are applied to each vertex $x_k \in S_{i+1}$. The vertex x_k is connected with the closest one in I'.

Steps 4 and 5 are repeated for all sets S_i sequentially until ST is built.

There are many modifications of these procedures. The vertex coordinates can be sorted not by increasing S (T), but by decreasing S (T). Time complexity of procedure 2 is $O(cn^2)$. Fig. 3.8 shows the implementation of procedure 2 with sorting of vertices in step 1 according to decreasing S. Total length of the ST is 22, and there are two SP.

To build ST one can apply different methods, as well as the branch-and-bound algorithm, depending on ART requirements.

3.3 Mathematical Model of Scheme Planarity and Its Flat Layout

Non-planar graph; maximally planar graph; modified contiguity matrices defining graphs; directed covering tree; inverse edge; arc.

After determining the connections list, i.e. construction of SCT with Prim or Steiner principles, the problems of planarity must be solved. First, the CD graph model is analyzed. For this purpose, the set of all connected undirected graphs $M(G)$ is divided into three non-intersecting subsets M_1, M_2 and M_3. Subset M_3 includes all connected graphs meeting the condition

$$m \leq (n + 2). \tag{3.4}$$

The graph subset M_1 contains all certainly planar graphs. For the connected graphs, included in subset M_2, the number of vertices and edges should satisfy the following inequality

$$m > 3\,(n-2). \tag{3.5}$$

All graphs included in M_2 are certainly non-planar. It results from the fact that the maximum number of a planar graph edges is $3(n-2)$. The subset M_3 includes connected graphs with the number of edges satisfying

$$(n+2) < m < 3\,(n-2).$$

The subset M_3 includes both planar and non-planar graphs.

We shall now consider a method of determination of planarity for graphs from the subset M_3. The algorithm consists of three parts. The first part is devoted to building of a directed graph D from a source graph G after a depth search. The directed graph D is simpler than the source graph, as it results from deletion of paths and cycles. The second part considers questions connected with the search for cycles in graph D. The third part concerns construction of a planar graph by inverting the edges that correspond to previously found cycles.

Assume an undirected simple connected graph G without loops with n edges. Graph G is described by matrix $\mathbf{R} = [r_{i,j}]\ n \times \rho$, where ρ is the biggest local degree of graph vertex. Element $r_{i,j} = n$ if edge (i, n) exists in the graph. The tree, built through depth search in source graph G, is called directed covering tree (DCT). An edge in a directed graph $D = (V, W)$ is an inverse edge, if there is a path in DCT from vertex w to vertex v and the edges of graph D that are not in DCT, are inverse edges.

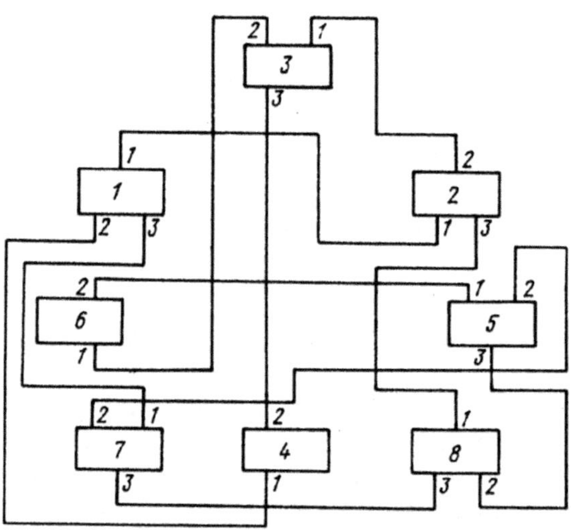

Fig. 3.9. A fragment of circuit diagram

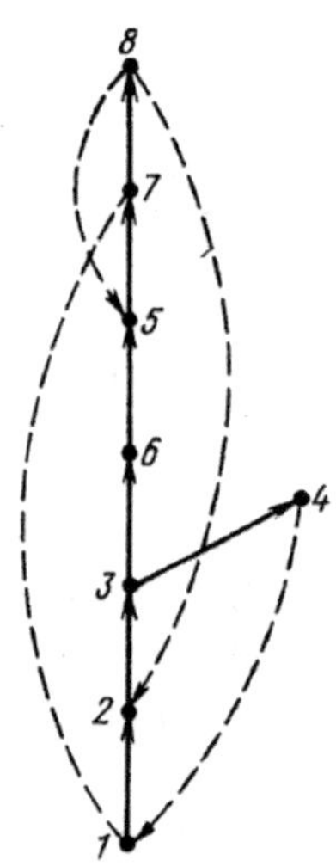

Fig. 3.10. Directed graph for CD fragment of Fig. 3.9

As an example we will examine a CD fragment of Fig. 3.9; an appropriate modification of its adjacency matrix is shown below.

$$\mathbf{R} = \overset{\textit{Contacts}}{\left|\begin{matrix} 2 & 4 & 7 \\ 1 & 3 & 8 \\ 2 & 6 & 4 \\ 1 & 3 & 0 \\ 6 & 7 & 8 \\ 3 & 5 & 0 \\ 1 & 5 & 8 \\ 2 & 5 & 7 \end{matrix}\right\|} \textit{Elements}$$

The first part of the algorithm consists in obtaining matrix $\mathbf{B} = [b_{i,\,j}]$, describing a directed graph D. Graph D is built from undirected simple connected graph G, represented by matrix $\mathbf{R}$, using depth search. The algorithm of construction of the directed graph D is shown below.

Let v_i be the number of vertex that is examined at current moment of time. The algorithm is as follows:

1. Start from vertex v_i, $i = 1, n$.
2. By passing over the edge (v_i, v_j), $i \neq j$, incident to examined vertex v_i, the next vertex v_{i+1} is examined.
3. It is established whether v_j is a new vertex (or it was accounted for before), i.e. whether the edge (v_i, v_j) belongs to DCT or not. If not, go to step 4, otherwise go to step 5.
4. Edge (v_i, v_j) is an arc in DCT. Take the new vertex v_j as the current one, i.e. $v_i = v_j$, then go to step 2.
5. The edge (v_i, v_j) that was used to pass to v_j, is an inverse edge.
6. Check whether the examined vertex v_i has another incident edge that was not considered before. If there is one, go to 2, else go to 7.
7. Check whether graph is D entirely built, i.e. all edges of the source graph G were examined. If not, go to 8, else go to 9.
8. Return to the previously examined vertex v_{i-1} via the DCT arc that was used to pass to the currently examined vertex v_i. Vertex v_{i-1} becomes the currently examined vertex ($v_i{:} = v_{i-1}$); go to 6.
9. Algorithm terminates work. Construction of directed graph D is complete.

Matrix $\mathbf{B}$ has the same dimensions as $\mathbf{R}$. Element $b_{i,j} = k$, if arc (i,k) belongs to DCT; and $b_{i,j} = k$ if arc (i,k) is an inverse edge. First row of matrix $\mathbf{R}$ is examined, element of matrix $\mathbf{B}$ is set to the value of the first element of matrix $\mathbf{B}$ different from zero, located in this line. So, process starts from first vertex, and a pass via arc (1,*k*) is made to vertex *k*, as *k* was never visited before. Sequences *S*(*L*) are formed, $L=\overline{1,n}$, composed of the numbers of examined vertices *C* (*L*), $L=\overline{1,n}$, and numbers of columns with the examined vertices in matrix **R**, and *SP(L)*, which include vertices in the order of DCT construction. The sequences *S*(*L*) and *C*(*L*), $L=\overline{1,n}$, are used when a return is necessary to an already passed vertex, to continue examination of edges, incident to it. The sequence *SP(L)* is used in the second part of the algorithm for building of the matrix that describes cycles in the directed graph *D*. The operations are repeated as described above, until all edges of the source graph are examined, i.e. graph *D* is built (Fig. 3.10). Graph *D* is described by the matrix **B**, shown below:

$$\mathbf{B} = \begin{array}{c|ccc|}
1 & 2 & 0 & 0 \\
2 & 0 & 3 & 0 \\
3 & 0 & 6 & 4 \\
4 & -1 & 0 & 0 \\
5 & 0 & 7 & 0 \\
6 & 0 & 5 & 0 \\
7 & -1 & 0 & 8 \\
8 & -2 & -5 & 0
\end{array}.$$

The second part of the algorithm consists in determination of all cycles in graph *D* and construction of matrix $\mathbf{R}_c$, describing those cycles. The number of cycles in graph *D* is equal to the number of inverse edges in it, i.e. to the number of negative elements in matrix **B**, as all cycles in *D* consist of an empty path in DCT and an inverse edge. Determination of cycles in *D* is made in the following way:

First, the cycles are found, whose corresponding inverse edges begin in a vertex that was examined as last during construction of DCT; then the cycles, whose inverse edges begin in the vertex, passed next to last during construction of DCT, and so on. Cycle search begins from the end vertex *v* of the inverse edge, corresponding to it. When passing over the DCT arc (*v*, *w*) a vertex *w* is visited. If this vertex is not the beginning of some inverse edge, then we pass through an arc, incident to it, and then over to the next vertex, and so on. If vertex *w* is the beginning of some inverse edge, then it is checked, whether its final vertex is the one, from which the search began. If yes, then a cycle is found; if not, then the next arc, incident to *w*, is examined. When all arcs, incident to vertex *w*, have been examined, and a cycle has not been found, return via arc (*u*, *w*), that was used to pass to *w*, and continue to move by another arc, incident to *u*. This is done until we come to an inverse edge, corresponding to a cycle sought.

We introduce matrix $\mathbf{R}_c = [dM_i, M_j]$, $M{\times}n$, where *M* is the number of inverse edges in *D*, *n* is the number of vertices included in the cycle of maximum length, i.e. the number of vertices in the graph. It is known that $M = m{-}n{+}1$, where *m* is the number of vertices in the initial graph *G*.

In the general case the maximum number of inverse edges in a planar directed graph D is $2n$–5. To make a directed graph planar, the number of its edges should be less or equal ($3n$–6). This number results from the number of edges of the covering tree and the number of inverse edges. So, the maximum number of inverse edges is $3n-6-n+1 = 2n-5$.

Rows of matrix $\mathbf{B}_c$ contain, consecutively, all vertices that form cycles in D, i.e. each row contains the ordered set of vertices that form one separate cycle. First column of this matrix is formed by the vertices that are the last vertices of inverse edges, and the last column - by the first vertices of the inverse edges.

The third part of the algorithm consists in building of planar graph D by rearrangement of cycles, i.e. of the inverse edges, corresponding to them. As D consists of DCT and reverse edges, and every cycle in it consists of DCT edges and one inverse edge, planar graph construction reduces to arrangement of inverse edges in the directed graph without intersections.

Assume that inverse edges can be placed in two planes relative to DCT. One of them is marked with "+", and the other with "–". Assume that an inverse edge, starting from the vertex that was passed as the last during the main tree building process; and the cycle, corresponding to it, shown in the first row of matrix $\mathbf{R}_c$, are in the "positive" plane. Try to place in this plane the cycle from the second row of $\mathbf{R}_c$. The algorithm of placing a new cycle in one plane with a one previously arranged is shown below.

1. If new cycle has a vertex that is the final vertex of any previously arranged cycle inverse edge, go to step 2, else go to 6.
2. If the initial vertex of this previously arranged inverse edge belongs to the new cycle, then go to step 6, else go to 3.
3. If the second vertex of the previously arranged cycle belongs to the new one, go to 4, else go to 6.
4. If the initial vertex of the inverse edge arranged is a ramification in the covering tree, i.e. it has local degree bigger than 1, then go to 6, otherwise go to 5.
5. The previously arranged cycle is moved to another plane. The signs of elements in the row of matrix R_c, corresponding to this moved cycle, are changed to the opposite ones, i.e. from "+" to "–" and from "–" to "+".
6. The new cycle is placed in the same plane with the previously arranged cycles.

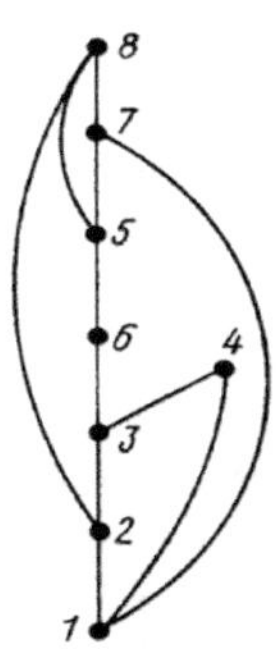

Fig. 3.11. Planar graph of CD fragment (Fig. 3.9)

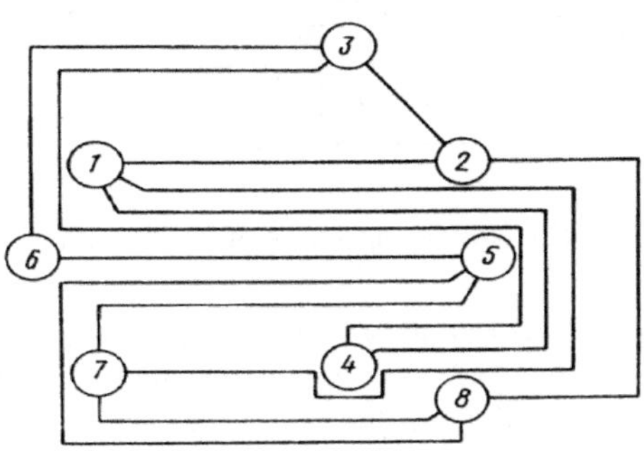

Fig. 3.12. Flat layout of CD graph of Fig. 3.9

If the cycle is located in the "positive" plane, together with the cycles previously arranged in it, the next cycle is examined. If not, the previously arranged cycle is moved to negative plane and it is checked in the same way, whether it can be placed with the previously arranged cycles. If not, then the previously arranged cycle is moved to positive plane, the signs of elements in the corresponding row of matrix $\mathbf{R}_c$ are changed from "–" to "+". Then, in a similar way the possibility is checked of placing a cycle, moved to the positive plane, with cycles previously arranged in it. The process continues thereafter as described above. If a cycle, moved to the positive plane, makes the initial cycle move to the negative one, then the graph is not planar. The process will continue until it is obvious that the graph is not planar, or until all inverse edges in D will be arranged without crossings, i.e. until a planar graph is built (Fig. 3.11), described, for the example considered, by matrix R_c', shown below:

$$R_c' = \begin{Vmatrix} -2 & -3 & -6 & -5 & -7 & 0 & 0 & -8 \\ -5 & -7 & 0 & 0 & 0 & 0 & 0 & -8 \\ 1 & 2 & 3 & 6 & 5 & 0 & 0 & 7 \\ 1 & 2 & 3 & 0 & 0 & 0 & 0 & 4 \end{Vmatrix}$$

It is necessary to note that in general case, for a required fixed allocation of elements, after flat graph layout is obtained, the total length of connections can increase. That is why it is advisable to combine the processes of planar graph determination for the scheme and allocation of its vertices, with the common criterion of the total length of edges.

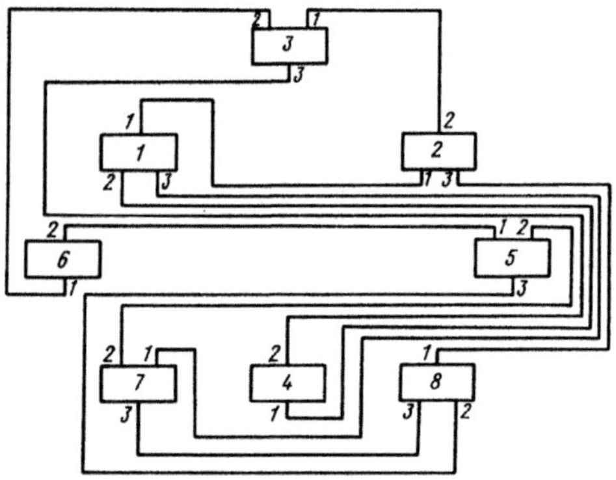

Fig. 3.13. Effect of the operation of the algorithm for CD of Fig. 3.9

The algorithm allows, upon examining the already built matrix representation of the initial graph, to obtain the matrix defining oriented graph, with a simpler structure than that of the initial graph. This helps in finding all cycles in it and, by moving corresponding inverse edges, in building the planar graph. So, for the example examined we have flat layout of CD graph (Fig. 3.12) and CD will look like in Fig. 3.13.

3.4 Distributing Connections in Layers

Via interconnections; striping; covering rectangle; incompatibility graph; combined topology; bichromatic subgraph; orthogonal routing.

After the stage of determination of planarity, when CD graph is not planar, the problem assignment of graph or hyper-graph edges (CD connections) to layers is solved. The goal is the most effective use of CF with simultaneous optimization of CD construction parameters, like the number of layers, via interconnections, percent share of realized connections.

Depending on organization of routing process stripping is performed either before, or after, or during routing.

The main idea of the *stripping algorithms*, which work before routing, is in a priori isolation of conflicting groups in CD that cannot be assigned to one layer because of unavoidable crossings.

An often used method for striping before routing is a simple algorithm of forming two subsets of connections, divided in orthogonal coordinates in two layers. The subsets of vertical and horizontal segments are assigned to different layers. Solution is obtained fast, but, obviously, it is not optimal from the point of view of the routing area and the number of via intersections (Fig. 3.14).

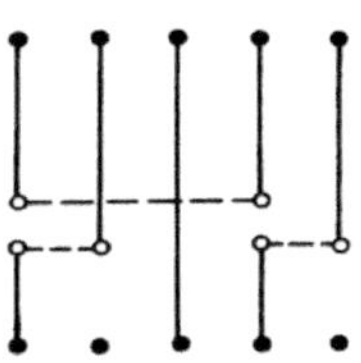

Fig. 3.14. Example of two-layer connection construction

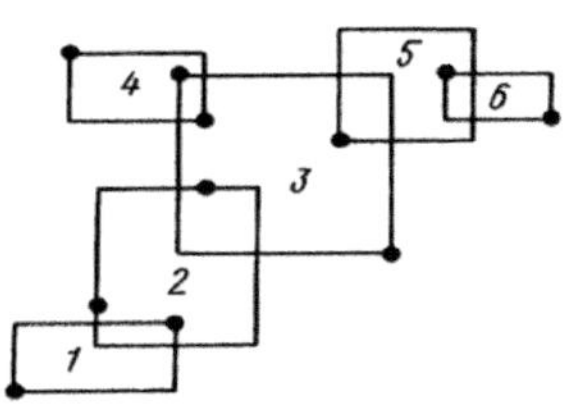

Fig. 3.15. Example of covering rectangles construction

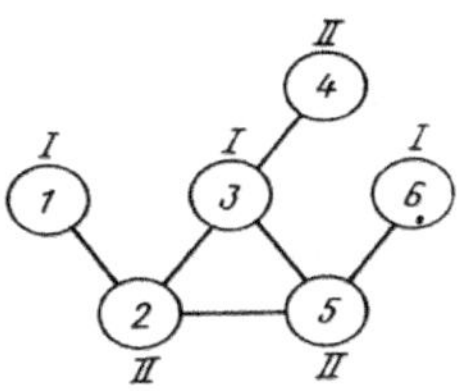

Fig. 3.16. Overlap graph for the example of Fig. 3.15

It is advised to set an "interaction" degree of connections during solving of the striping problem by building covering rectangle R_i for each circuit e_i. It is considered that two circuits overlap, if $R_i \cap R_j \neq \varnothing$. An overlap graph $G_p = (E, U)$ is built, where vertices $e \in E$ correspond to circuits, and edges $u \in U$ - to overlaps between them.

It is obvious, that the chromatic number γ_p of the graph $G = (E, U)$ is the upper bound on the number of layers, required to realize all CD connections. The value of γ_p is called the degree of overlapping of the system of subsets e_i. Naturally, all connecting trees for e_i can be assigned to one layer when $\gamma_p = 1$.

Let us examine an example of construction of covering rectangles for *G*-circuits (Fig. 3.15). For this example it is easy to build the overlap graph G_p (Fig. 3.16). According to the algorithm we can color graph G_p with two colors (I and II in Fig. 3.16). Then we have $\gamma_p = 2$. In other words, the scheme can be realized with two layers and it has two non-crossing subsets: $\{e_1, e_3, e_6\}$ - first layer; $\{e_2, e_4, e_5\}$ - second layer.

Calculation of geometric limitations for this procedure can be expanded by taking into account electric and other parameters during realization of circuits e_i and e_j in one layer. In such case, additive function is calculated:

$$F_{i,j} = \sum_{i=1}^{n} \alpha_{i,j} p_{i,j} ,$$

where $\alpha_{i,j}$ are heuristic weight coefficients, and $p_{i,j}$ are given parameters of the circuits.

Here, as in the previous case, the incompatibility graph $G_p = (E, U)$ is built, with, however, weight $F_{i,j}$ assigned now to each $u_{i,j} \in U$. The striping problem is reduced to construction of such a graph coloring with γ colors that the sum of weights of edges connecting vertices of same color is minimal. This coloring can be done only by one of known algorithms.

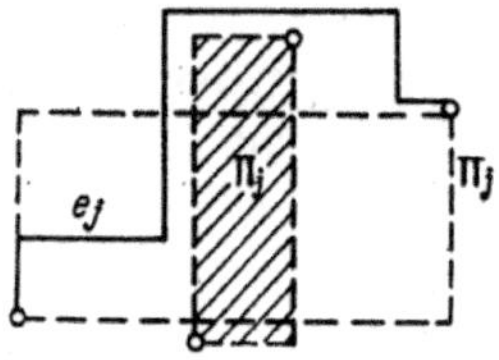

Fig. 3.17. Example of circuit construction

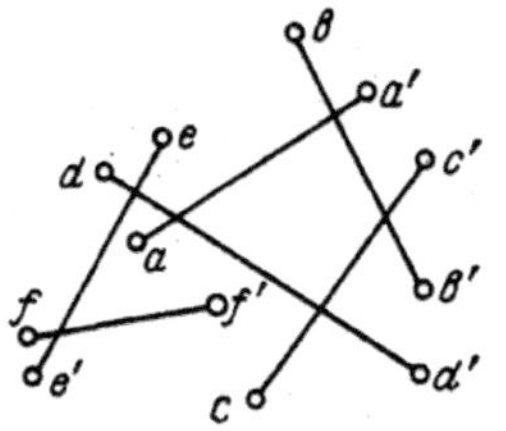

Fig. 3.18. Placement of the set of conductors on a plane

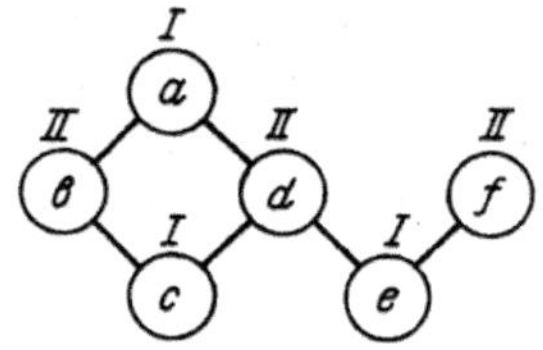

Fig. 3.19. Overlap graph for the set of conductors (Fig. 3.18)

We notice that the degree of influence of e_i on e_j can be estimated as the number of shortest paths for connecting e_j through R_j. In Fig. 3.17 it is shown how the shaded inadmissible area prevents from optimal realization of circuit e_j.

Let us now examine how the striping process is organized, if it is carried out after routing, when the combined topological draft is available.

The theoretical basis of this method was developed by Codres, Weisman and Bader. We analyze the statement of the striping problem on an example. Assume the set of conductors $M = (a, a'), (b, b'), (c, c'), (d, d'), (e, e'), (f, f')$ located on a plane as in Fig. 3.18. These locations correspond to the overlap graph $G_p = (E, U)$ (Fig. 3.19), where nodes $e_i \in E$ correspond to conductors, and edges $u \in U$ - to their crossings. The problem is reduced to finding the two-color coloring of G_p, with minimum total number of edges connecting vertices of the same color vertices. This is equivalent to isolation, in graph G_p, of a bichromatic subgraph with the maximum number of edges. On the basis of known property that a graph is bichromatic, if and only if it has no cycles of odd length, coloring procedure is applied, based on solving a linear programming problem. In our case graph is bichromatic, and the striping is: I — {$(a, a'), (c, c'), (e, e')$} and II — {$(b, b'), (d, d'), (f, f')$}.

Considering that as of now there are no criteria of isolating maximal n-chromatic subgraphs in the source graph, the methods considered cannot be applied to n-layer systems. That is why we use heuristic procedures of coloring in e colors, followed by sequential filling of layers with other connections.

Let us present an algorithm solving this problem for unlimited number of layers. Let there be given, for each circuit e_i, some variants of realization of the connection tree $\{D_i^j / j = 1, 2 \ldots n\}$ (in the simplest case $n = 1$). The peculiarity of this problem lies in the fact that for each pair of contacts, for which $n > 1$, one connection should be selected, so as to minimize the required number of layers.

Solution is obtained in two stages: in the first stage the maximal internal stable sets of the overlap graph G_p are defined, and in the second one its minimal and quasi-minimal coloring is established.

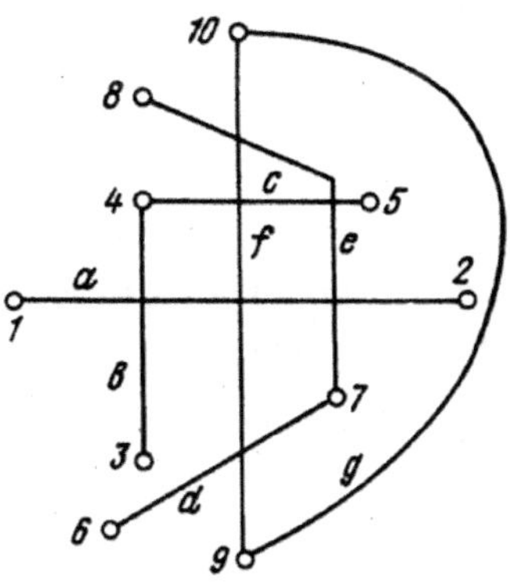

Fig. 3.20. Example of a set of conductors

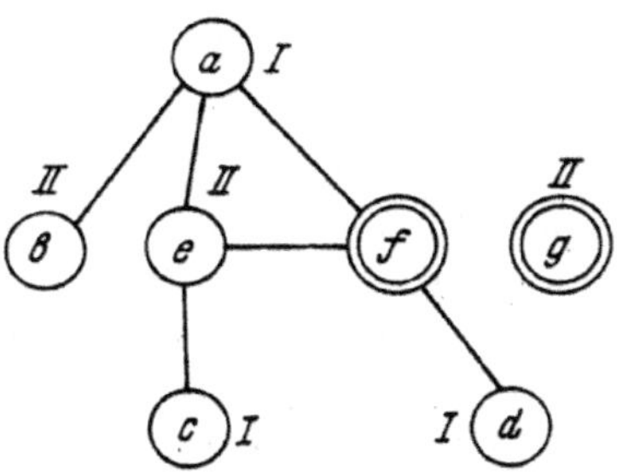

Fig. 3.21. Overlap graph for the example of Fig. 3.20

An example is shown in Fig. 3.20. After having applied the algorithm to determine the family of independent sets we obtain for G_p (Fig. 3.21) the following maximal independent sets (IS):

$\varphi_1 = \{g, a, c, d\}$, $\varphi_3 = \{g, f, b\}$,

$\varphi_2 = \{g, b, e, d\}$, $\varphi_4 = \{g, b, d, c\}$.

Having the family $\psi = \{\varphi_1, \varphi_2, \varphi_3, \varphi_4\}$, we pass to second stage, connected with finding of minimal, quasi-minimal or locally optimal coloring of vertices of the overlap graph G_p. Depending on the complexity of CD the algorithms used are either precise (ART = $O(n!)$), iterative (ART = $O(n^2)$) or sequential (ART = $O(n)$).

Solving this problem can be reduced to finding of the vertex set covering in graph G_p with minimal number of independent subsets.

We select subsets in E, corresponding to possible realizations of each connection $\{a\}$, $\{b\}$, $\{c\}$, $\{d\}$, $\{e\}$, $\{f, g\}$. This information will be taken into account while selecting one of connection variants in covering subsets.

It is obvious that the initial graph G_p is covered by two IS, {φ1} and {φ2}. Considering the connection possibilities, *g* or *f*, for realization of contacts of 9-10 we select connection *g* and, by removing concurrent connections from the covering IS, we obtain striping in two layers: I: {*a*,*c*,*d*} and II: {*g*, *e*, *b*}. It should be noticed that if only one connection were examined for contacts 9, 10, then three layers would be required to solve the problem.

Here is an heuristic algorithm that allows reducing via intersections for orthogonal striping.

1. Select all non-crossing circuit segments and mark them. (Obviously, they can be assigned to any layer without removing via intersections.)
2. Select in the set of unmarked segments the subset, which has a segment adjacent to at least one marked circuit segment.
3. Put each segment $z \in Z'$ on the layer, where the marked segments are placed, adjacent to it. If there are several adjacent segments, select the one that requires the least number of via intersections.
4. Mark the allocated segment.
5. If $z' = 0$, then go to 6, else go to 3.
6. End.

Here, ART equals $O(n^2)$! It is important to note that selection of the striping algorithm is a nontrivial problem, requiring thorough analysis of initial requirements. For striping during routing, dynamic programming or sequential algorithms with estimation of each step and return back, if the new result is worse than the previous one, are mainly used.

3.5 Routing Algorithms

Wave front; wave propagation; method of conditional optimization; left edge algorithms; ray; horizontal, vertical channels; trunk; vertical limitations graph.

In practice, sequential routing algorithms are commonly used, and, among them, especially the wave algorithm. In addition to the disadvantages of all the sequential routing algorithms, the wave algorithm is connected with significant time costs and high computer memory requirements.

Here are the main principles of Lee's wave algorithm.

1. Routing plane (board, crystal) is divided into rectangular areas (discrete values) of predefined size. The discrete size of the area is defined by the admissible sizes of conductors and spaces between them. The problem of track laying is reduced to obtaining of a sequence of discrete values connecting elements x_i and x_j, corresponding to outlets (contacts) of the scheme elements.
2. The weight function $F = F(f_1, f_2 \ldots f_r)$ is introduced; it is the criterion of path quality; f_i characterizes a path in terms of length, number of crossings, via intersections, curves, etc.

3. To an element x_i and the discrete values, neighboring with the previously analyzed ones, a definite value of the weight function $F = m_i$ is assigned. Step 3 is iterated until element x_j obtains the weight value m_j.
4. Starting from element x_j, movement to element x_i is made, so that the discrete values of the weight function decrease monotonously. As a result, we have a track that connects elements x_i and x_j.

Let us examine the process of building track from x_i to x_j in detail. On the plane, where the track is to be built, wave propagation is modeled from the "source" x_i until wave front reaches the "receiver" x_j, or until, on a kth step, the wave front absorbs no free discrete values. Note that wave propagates only via free discrete values. This part of the algorithm is called "wave propagation". If after its finalization the element x_j is reached, the second part of the algorithm, called "track laying", is performed. During wave propagation the algorithm builds sequentially, step-by-step, 1st, 2nd… k^{th} front of source x_i.

A new front building process starts with assignment of weights m_{i+1} to discrete values, neighboring with the discrete values of the previous front. The weight $m(i_k,j_k)$, assigned to discrete values of the kth front, is per unit greater, than the weight of discrete values of the $(k-1)^{th}$ front. Besides the weight m_k, to each discrete value, a path coordinate is assigned that defines the discrete value, from which at this moment the front arrived. As the track building process follows the path coordinates, step 4 of the algorithm is performed.

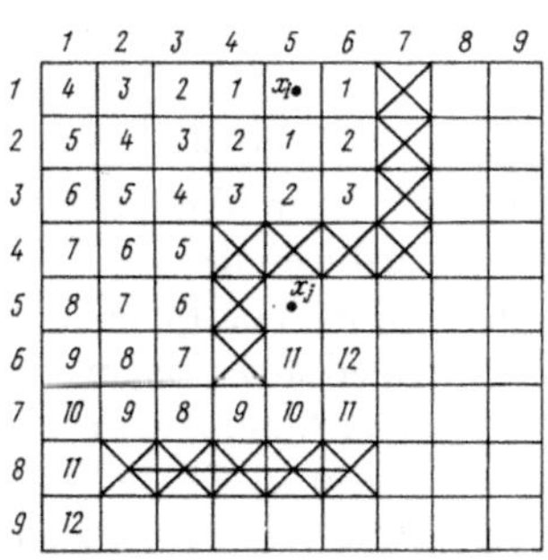

Fig. 3.22. Example of connection of cells xi and xj using wave algorithm

Fig. 3.22 shows an example of connection of elements x_i and x_j by the wave algorithm. Note that there are several variants of path building, from which computer or constructor selects one that compiles the best with the predefined criteria.

To improve the performance of the wave algorithms it is recommended:

- to spread simultaneously wave from two points being connected;
- to begin routing from the farthest point from the centre;
- to reduce the routing area by using the pair of contacts being connected to form a rectangle equivalent to a new artificial border.

To decrease the required computer memory, it is advised to fulfill the following condition: the predecessor labels should be different from the receiver labels. This allows for using the coding sequence 1—1—2—2—1—1—2—2—1... So, for definition of any front cell state, two bits of memory suffice. To reduce the number of path turns on the laying stage the following priority rule can be used. When moving from cell x_i to the front cell x_{i-1}, the direction should be kept, if possible, defined by the passage from cell x_{i+1} to cell x_i.

In the example of Fig. 3.22 the required number of storage bits per one CF cell is $\{log_2 (L + 2)\}$, where L is the maximum path length on CF.

The structure of Lee's algorithm allows for optimizing connections for different criteria, so we will examine some modifications that allow for performing connection optimization with a set of parameters.

For this purpose, an integral criterion of optimal path is used, in, for example, the form of the following additive function:

$$F = \sum_{i=1}^{n} \alpha_i f_i , \tag{3.6}$$

where α_i are a priori established weight coefficients that define importance of parameters f_i (they can be based on expert assessments). The weight of a free cell x_i, neighboring upon the front cell F_{k-1}, can be calculated using the following formula:

$$P(x_i) = P(x_{i-1}) + \sum_{i=1}^{n} \alpha_i \mu_i , \tag{3.7}$$

where $\mu_i \in \{0, 1\}$ is the sign of ith parameter change, and n is the number of parameters considered. From the set of the possible values $P(x_i)$ the minimal one is selected, and the following front includes free cells with minimal weight.

The method of parallel path optimization has disadvantages: the coefficients of "importance", α_i, of separate criteria cannot be always estimated and even when there are two-three parameters, the time of realization of the algorithm increases.

Because of this, the method of conditional optimization is used. A CF cell can be occupied, free or half-occupied. A horizontal (↔) or vertical (↕) conductor comes through a half-occupied cell. Intersections are allowed at right angles, and this makes a half-occupied cell occupied. The algorithm uses four lists S_1, S_2, S_3, S_4, and the cells of CF can be in eight states: occupied, free, half-occupied (↔), half-occupied (↕), or bearing one of the marks: ↑, →, ↓, ←.

The cells, put on lists S_1, S_2, S_3, S_4 are assigned index I, whose value is equal to the distance of these cells from the source cell x_i. Before the start of the algorithm, lists S_2, S_3, S_4 are empty, and cell x_i with index value 0 is on list S_1. Fig. 3.23 shows an example of functioning of this algorithm.

During the design of two-layer printed circuit boards the iterative algorithms with limitations to wave propagation are often used. Let us examine the algorithm of Heller and Fisher, where routing is performed in seven stages. In the first stage the circuits are traced connecting the closest contacts of one vertical column. Only

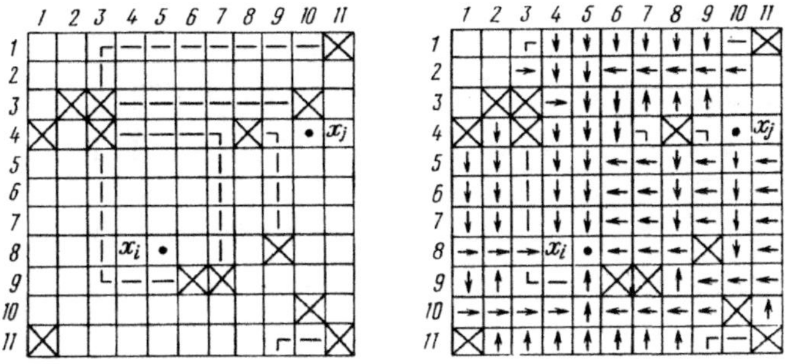

Fig. 3.23. Example of functioning of the algorithm (→ denotes the way found)

connections that are placed successfully in the vertical layer are fixed. For this, the discrete values around each outlet are reserved. In the second stage, connections in the horizontal layer are traced in the same way. In the third stage, connections are traced in the vertical layer having contacts in two neighboring columns. In the fourth - connections that have invariant external outlets are placed. In the fifth, using common wave procedure other connections that have external outlets are traced, and in the sixth, the remaining connections without external outlets are traced. After the sixth stage rerouting of connections, allocated during first three stages, is made, using previously reserved discrete values. In the seventh stage attempt is made to place remaining connections using discrete values that become free after rerouting.

In some algorithms the ideas are used of maximum flow determination in networks for conductor routing on PCB layers. For this, the respective board areas are replaced with a network, i.e. cells (areas) of the grid graph G_r are associated with network nodes, borders of areas of non-zero length - with edges connecting these nodes. In the preliminary stage the signs of interdiction and availability are put in respective nodes and edges of the network. The interdictions and delays as to passing over the edges correspond to prohibitions and limitations to track placement. This enables accounting for appearance of spurious capacitance during routing process.

The here examined modifications of the wave algorithm provide for construction of the minimum length path between points on the plane. Note that the use of the wave algorithm entails significant time costs for solving routing problems. Of this, wave propagation accounts for about 90% of all calculations, and path building for 10%. In some practical ECE schemes most connections have simple shape, and it is not required to examine all grid cells for path building. Consequently, it is reasonable to use the ray algorithm. The gist of the algorithm is to examine the grid and determine the path between two vertices x_i and x_j according to some predefined directions that are like rays. Before the start of the algorithm, a number of rays, spreading from areas x_i and x_j, are defined, along with the ordering of path coordinate assignment. Usually, the number of rays for each area (source) is set to two. The deployment of rays is done simultaneously from both sources until two opposite rays meet in some area.

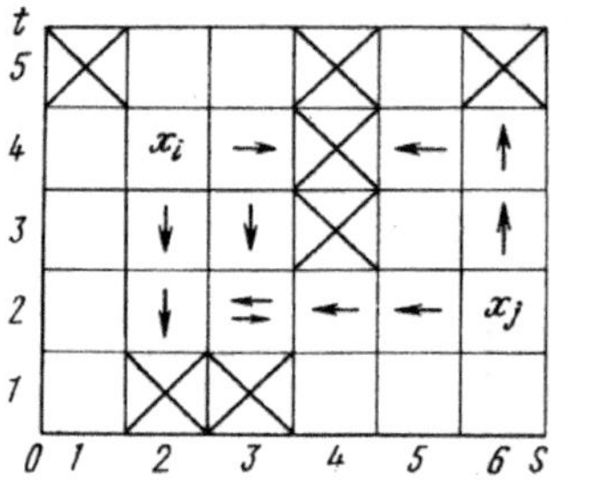

Fig. 3.24. Example of connection of cells x_i and x_j using the ray algorithm

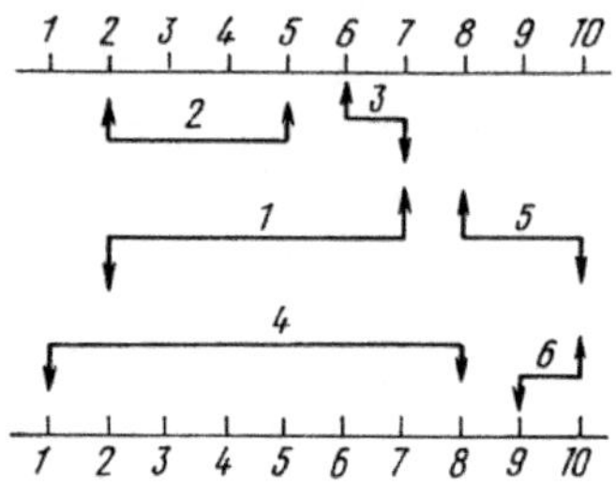

Fig. 3.25. Example for the channel routing problem

Fig. 3.24 shows an example of track building for connecting areas x_i and x_j using ray algorithm. For each of the sources x_i and x_j two rays with opposite directions are taken: $x_i^{(1)}$: down right, $x_j^{(1)}$: up left, $x_i^{(2)}$: right down, $x_j^{(2)}$: left up. If x_j were situated not to the left of x_i, but to the right, then path coordinates, "left" and "right", would have to be interchanged. After the first step of the algorithm, areas with coordinates (t_3; s_2), (t_2; s_5) are occupied. In the third step the rays $x_j^{(2)}$ and $x_i^{(1)}$ meet. As the ray algorithms do not always give a solution, it is purposeful to use them together with the wave algorithm. Usually, the use of the wave algorithm allows for building of 60% of tracks, the remaining tracks being built using other algorithms or by the constructor. The ray algorithms are mostly advised for use in routing of boards with small degree of cell filling. In this case they allow for a significant time saving in comparison with the wave algorithm.

Recently, channel routing algorithms have gained wide popularity. They are based on track building by enlarged discrete values of TF, for which a system of horizontal and vertical channels is used. Channel width depends on the number of connections, allocated in it. A channel is divided into lines, called trunks. Any connection in channel routing will consist of trunk parts, united into one chain. Realization of the channel algorithm assumes performing two procedures: distribution of connections by channels considering their optimal load, and optimization of connection allocation in channel trunks. For a definite construction-technological basis of PCB production, each channel of the wiring plane is assigned a number, called channel capacity. It is the maximum allowed number of conductors passing through the channel section, taking into account technological limits. In the simple cases the procedure of optimal connection distribution between channels leads to their uniform load, while in more complex cases, additionally, calculation of electromagnetic and thermal compatibility of neighboring conductors is performed. The purpose of the second procedure, optimization of connection allocation in channel trunks, is to minimize the number of via between different layers of wiring plane.

The problems of the first procedure are solved using the SCT algorithms. In construction of the SCT for each connection, the load of every channel is considered that it passes through. After SCT has been built for every circuit, the load z_i is determined for each channel and compared to the corresponding channel capacity

Y_i. If $z_i > Y_i$, then m circuits, where $m = z_i - Y_i$, are to be retraced, using only channels having $z_j < Y_i$.

Another approach assumes calculation of current channel capacity Y_i, i.e. the number of conductors that can (still) be allocated to this channel. A channel "closes" for connection allocation as soon as $Y_i = 0$.

Wave procedures are often used for connection distribution between channels, instead of SCT determination, with channels represented by discrete values for wave propagation on CF.

The problem of final routing is commonly formulated in the following way. We have a horizontal channel, limited by the top and bottom contact rows PT_i and PT_j, where $i, j = 1, 2, \ldots$ Between the top and bottom rows, in the sets of free lines, trunks are defined: $S = \{S_k/k = 1, 2 \ldots m\}$, where m is channel width. We must connect all abscissas of the same label contact groups (each group corresponding to one circuit) with commonly broken lines, coming through trunks parts; and afterwards - connect contacts with vertical segments. Routing should be done according to a selected quality criterion (overall length, number of via intersections, realized connections, occupied trunks, etc.).

The generic statement of and the procedure for the channel routing problem for a horizontal channel can be illustrated for four-sided channel.

The first channel routing algorithm (the so called "classical" problem statement that does not allow sharp curves or transferring horizontal connection segments among trunks) was suggested by Hashimoto and Stevenson in 1971. This algorithm is also called the "left edge algorithm".

The example used was given in Fig. 3.25 with initial routing specifications.

The initial data are defined by connections list

$$\mathbf{R}_L = \begin{pmatrix} 0 & 2 & 0 & 0 & 2 & 3 & 1 & 5 & 0 & 6 \\ 4 & 1 & 0 & 0 & 0 & 0 & 3 & 4 & 6 & 5 \\ 1 & 2 & 3 & 4 & 5 & 6 & 7 & 8 & 9 & 10 \end{pmatrix},$$

where the first row defines top contact row, the second row - bottom contact row, and the third - contact numbers.

Contacts, marked with zeros in the connections list are free or cannot be connected. Contacts with a non-zero number i are to be connected with circuit i.

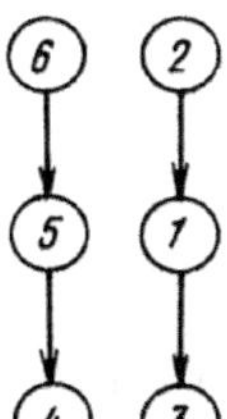

Fig. 3.26. VLG for the example of Fig. 3.25

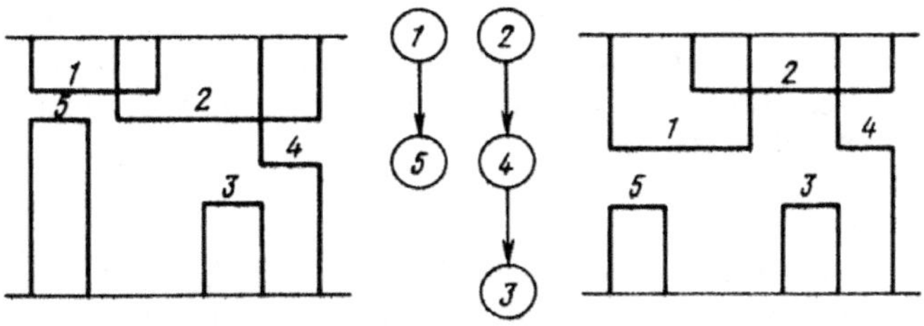

Fig. 3.27. Solution obtained using the left edge algorithm (four trunks are required) (a), and the optimal solution (three trunks are required) (b)

As intersections of vertical circuit segments should not be allowed, we must introduce a respective constraint (i.e. relations between TF circuits). Such relations are usually represented by the vertical limits graph (VLG), that is, a directed graph $D=(X,\vec{U})$, where X corresponds to the set of connections, and any $a\in X$ and $b\in X$ are connected with an arc from a to b, if the constraint of inadmissibility of the vertical segments intersection should apply. For the example of Fig. 3.25 the VLG of two connectivity components is shown in Fig. 3.26.

Vertex i is called a parent of vertex j (j is a child of i) when an arc in VLG is directed from i to j. We introduce also the concept of density P_i, being the number of horizontal connection segments crossing a vertical channel i.

The main idea of the left edge algorithms is as follows. Horizontal trunks are sequentially processed, starting from the bottom and ending with the top. In a step, horizontal segment with the least abscissa among all segments not having parents in VLG, is allocated to the trunk. This process continues until the current trunk is filled without violation of limits. We then pass over to the next trunk and repeat the process. It is easy to see that the left edge algorithm does not always give optimal solution. Thus, Fig. 3.27 shown the solution obtained using the left edge algorithm, and the optimal one.

So, when choosing a trunk, it is advised to consider not only such connection characteristic as the left end coordinate, but also connection length, length of path in VLG, where connection is allocated (in Fig. 3.27 connection 1 is ought to be the first, so as to shorten the limiting chain), density of columns, crossed by connections, etc.

We can, therefore, propose the following modification of the left edge algorithm:

1. Create VLG and calculate column density. Let N be the set of segments, and $T=1$, trunk counter.
2. While $N=\varnothing$, execute algorithm, else go to 8.
3. Form N_0, the set of segments that have no parents in VLG.
4. Find $N_a \subseteq N_0$, such that:

   There are no two crossing segments in N_a;

   2) $\sum_{n\in N_a} W_n \to \max$, where W_n is weight of segment n.
5. Allocate segments from N_a to trunk T. Form $N=N\setminus N_a$.
6. Reconstruct VLG considering the allocated VLG segments.
7. $T=T+1$. Go to 1.
8. End.

This algorithm differs from the left edge algorithm only by step 4, where the connection weight accounts not only for the left end coordinate of the connection, but also for its other characteristics. Here is the following additive function, defining it:

$$W_n = \sum_{z \in Z_i} U_z + (\alpha \rho_i + L_i)|z_i|,$$

where z is the number of columns crossed by connection i; U_z is density of column z; L_i is length of maximum path in VLG; ρ_i is degree of vertex i in VLG; α is a coefficient (is set to 3). The ACC of the procedure is $O(\beta n^2)$.

Note that the presence of cycles in the VLG can cause a dead-end situation for the classical problem statement.

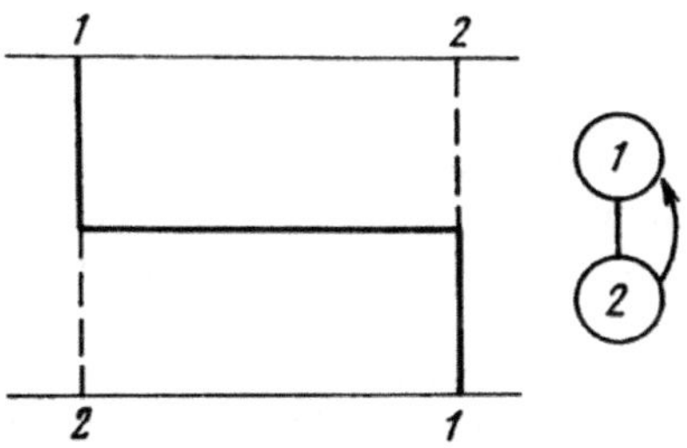

Fig. 3.28. Example of a dead-end situation

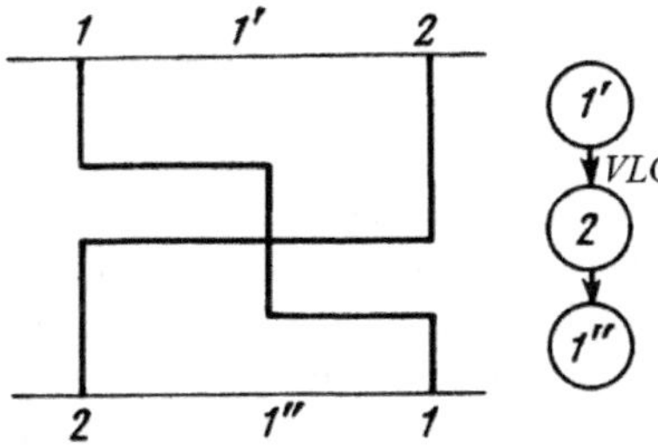

Fig. 3.29. Example of use of pseudo-contacts

Let us examine the example of Fig. 3.28. The problem cannot be solved for the classical statement, and so it is necessary to use "doglegging" or breaks of horizontal segments from trunk to trunk. In such cases additional pseudo-contacts (1' and 1'' in Fig. 3.29) are used, and near them connections are broken. After that we can operate with segments 1-1' and 1''-1 in a normal manner [84].

There is a routing system for irregular structures, based on channel algorithms. The irregularly allocated components are grouped in macro-cells. The free space between macro-cells is divided into horizontal and vertical channels. Routing is performed in three stages:

1. Inside every macro-cell;
2. Preliminarily, between the cells;
3. Inside channels (using the modified left-edge algorithm and broken lines for elimination of the dead-end situations).

Note that channel algorithms have higher performance and lower computer memory requirements than wave algorithms, but they are less universal. This limits the set of constructive-technological problems that can be solved using these algorithms. The wave, ray, and to a lesser degree the channel algorithms are based on sequential routing principle, assuming optimal design of every separate track without a foresight about the following connections on wiring plane. The disadvantage of this principle becomes apparent by the end of the solution process, when it becomes obvious that some of the previously allocated tracks could be built in "non-optimal" way, leaving more resources for allocation of subsequent tracks.

The methods of non-automated design, used by experienced designers, resulted in creation of "flexible" routing algorithm (by R. P. Bazilevich, G. E. Shiro), characterized by the absence of strict fixation of built tracks. Strict fixation often gives rise to unjustified obstacles for tracks that have not been built yet.

The flexible routing method consists of two main stages:

1. Macro-routing (construction of an optimal topological model for all tracks, providing for their flat inclusion in DTWF);
2. Micro-routing (construction of geometrical models for each track).

In other words, in the first stage circuits are arranged in rather large areas, inside of which they can "float", securing optimal conditions for building other connections. In the second stage fixation of connections takes place within wiring space with definition of their geometrical characteristics.

For determination of topological models of individual circuits, special procedures have been developed, based on finding the SCT for CF. Sufficiently high performance can be ensured by the wave methods. In the SCT building process only topological obstacles are taken into account, and main metric parameters of CF (e.g. carrying capacity of narrow spaces) constitute limitations. The essential procedures on the macro-routing stage are propagation and reduction of wave, as well as modification and transformation of a topological model. This secures realization of various routing strategies. In routing of multi-layer structures it is possible to organize the procedure of three-dimensional wave propagation.

Note that the topological model sets bounds on track length, defined by the "passage", assigned for this purpose, which can vary during micro-routing from $l_{\min}$ to $l_{\max}$.

As indicated, in micro-routing stage the geometric characteristics of each track are defined: track configuration, overall number of all track turns, and also, for every track separately, exact coordinates of all characteristic track points, tracks width, etc. Constraints come in this case as information from the previous stage, where all track fragments are allocated by macro-cells. Micro-routing is done by optimal "inclusion" of topological models of tracks in two-dimensional Euclidean space.

In this context, tracks are usually classified in three main groups:

1. With arbitrary configuration;
2. Situated along two mutually orthogonal directions;
3. Realized in a macro-cells working field.

Allocation of conductors in flexible routing algorithms is done sequentially, but the previously allocated tracks are not strictly fixed. This creates favorable conditions for allocating subsequent connections. Besides, presentation of the wiring plane as a system of faces turns out convenient for constructions using large-scale elements.

The main disadvantage of flexible routing algorithm is its complexity (ACC ≈ $O(n^3)$—$O(n^4)$), leading to high computer memory and time requirements for problem solution.

3.6 Improvement of Composition, Allocation and Routing Results Based on the Method of Equivalent Logical Function Translation

Equivalent; logical function translation; sub-item; base function; isofunction; base function reassignment. In the majority of existing CAD systems, allocation of CD elements and connection routing is done not considering operation logic of the schemes themselves, and so the possibility of transformation of scheme topology remains unused. Any automation, defined by its input and output, can be realized in a variety of ways. Usually, the way of realization is not selected taking into account future allocation and routing. A developer solving the problems of composition, allocation and routing, does not bring in any changes into the automation realization process. Logic schemes transformation can be used for creation of improved algorithms for composition, allocation and routing, with consideration of the operational logic of the scheme.

Let us examine a method of building improved algorithms, consisting in formation of equivalent parts of the scheme, followed by reassignment of connections groups, belonging to these parts.

There are different forms of presenting logical functions, but they reflect the same input-output table. Presentation form affects the specific scheme of realization of each function, i.e. type of elementary logical components, their connection topology, time characteristics, etc. Logical functions are equivalent, if they realize the same input-output table. For example, following functions are equivalent: $A_1 = \overline{B_1 + B_2}$ and $A_2 = \overline{B_1} \cdot \overline{B_2}$, $A_3 = B_1(B_2 + B_3)$ and $A_3 = B_1 \cdot B_2 + B_1 \cdot B_3$. Any electric scheme can be represented as a finite deterministic automaton. It is possible to select in a scheme the subsets of isomorphic sub-automata. They can be, e.g., power paths of amplification, signal forming, or smaller circuit nodes like filters, amplifiers, registers and others, down to NAND or NOR elements.

Here are five main stages of improving the functioning of the algorithm:

1. Selection of isomorphic subsets of sub-automata;
2. Analysis of source information, selection of path and solution criteria;
3. Formation of solution space based on the method of equivalent translation of logical functions;
4. Selection of a pair of isomorphic automata and realization of translation of commutation field topology (one-to-one switching of connection groups belonging to the examined pair of sub-automata);
5. Evaluation and output of results.

Let us examine the method of forming solutions space based on the principle of equivalent logical function translation. Assume a set of elements $X = \{x_i, \mid i = 1\ldots n\}$ (chips, micro-assemblies, LSI components, etc.) and a scheme of their connection are given. In general case, any element x_i can be represented as a set of in dependent contact groups; and for each contact a separate elementary logical function can be realized. Such a contact group will be called sub-element. For example, the K155LA3 chip consists of four two-input sub-elements AND-NOT. So, each ith element of set X can be assigned a subset $X_i^* = \{x_{i,j} \mid j = 1, 2 \ldots\}$ of sub-elements $x_{i,j}$. Now we unite the subsets X_i^* into the sub-element set:

$$X^* = \{x_{i,j} \mid i = 1\ldots n; j = 1\ldots (j_i)_{\max}\}.$$

According to predefined initial conditions, a subset of elementary logical functions F_i is implemented with each element x_i. The functions, realized with definite elements will be called base functions. Most often, the value of a base function is congruent with the value of a logical function defining certain element. For example, the base function $f_q = \overline{A \cdot B \cdot C}$ is realized with sub-element 3NAND, etc. All subsets of base functions are united in the set of base functions F.

Each element X has some construction redundancy, if the number of elementary logical functions, realized on examined functions, is less than the number of sub-elements it consists of, or if the base function of sub-element is not maximal for this sub-element.

Now let us introduce the notion of isofunction. A logical function is called isofunction, if it is equivalent to a base function and is organized using construction-wise redundant sub-elements.

We form for each base function f_q a subset of all feasible isofunctions r_q^p, differing from each other as to realization complexity and weight P_q. The weight of an isofunction is the sum of information expenses (definition and use of variation capacity of input and output signals), calculation (possibility of reassignment, formation and re-establishment of the corrected connections table, etc.) and constructive (use of redundant construction elements, definition of CF space and introduction of new redundant elements and other CF topology modifications) resources.

If an isofunction cannot be realized with the examined set of elements X, its weight will be infinite ($P = \infty$). Base functions also can be seen as isofunctions with weight equal to zero ($P = 0$), as at the instant of examination they are already realized, and any other (even better than the current) variant of realization will require some expenses.

Establishment of isofunctions for any base function f_q is done in two stages: in the first stage the set of all possible abstract isofunctions is formed, in the second one, the isofunctions not implemented with the set X of source elements are removed from the set of abstract isofunctions. Thus, for the base function

$f_q = \overline{A \cdot B}$, realized with four-input conjunction NAND, 48 isofunctions (including base) can be identified, realized with 24 different sub-elements. The obtained set of the base function isofunctions includes only isofunctions realizable with elements of the initial set X^*.

So, for each base function f_q, a subset (family) of isofunctions f_q^p can be formed by applying certain predefined rules.
All isofunctions can be united to form the set

$$F^* = \bigcup_q \bigcup_p \bigcup_z f_q^p z. \tag{3.8}$$

The solution space of the problem considered is the set of assignments of functions from F^* to the set of sub-elements X^*.

Here is the statement and solution of the function re-assignment problem for the case when equivalent CD fragments do not intersect, for example, re-assignment of functions at the level of elementary logical functions, like NAND, NOR, implemented using finite deterministic sub-automata. Assume that there is a problem solution and some criterion K_w to evaluate the quality of this solution (the set of tracks T, the set of non-allocated connections T^* and the set of narrow spaces T^{**} are given). A narrow space is CF area, where distance between tracks, tracks and contacts, and dimensions of contact pads or tracks widths have definite bounds. The number of narrow spaces in CF is strictly regulated.

The variants of the base function reassignment can be selected depending on the routing criteria chosen for the problem, minimizing the number of non-traced connections, narrow spaces, and DWF border load deviation from the average for all discrete values of border load.

Here, the DWF border load value is understood as the value of the ratio of the number of tracks, passing through a given border, to carrying capacity of this border. In the improvement of routing, the criteria of the number of crossings and track turns are used as auxiliary ones.

In addition to these criteria it is advised to use the following rule: if some variants of reassignment have the same quality, the variant with the highest CF variation abilities should be selected.

Solution space is formed depending on the cardinality of the set of tracks T^* (that are to retrace under predefined criteria), available computing resources, envisaged design reserves and some additional requirements. In solution space forming process can take part as all elements from set X^*, as some part of them, selected by correspondent criteria.

In practically all modifications of the routing improvement algorithm, based on the method of equivalent translation of logical functions, it is necessary to divide the CF into micro-cells (discretes) (MD).

Let us divide CF into a set $M = \{m_i \mid i = 1\ldots B\}$ of MD. The MD configuration is established in interactive mode by the deciding person. A rectangular shape is assigned to each micro-discrete. The borders of MD should be defined by channels. Crossing of an element by the MD border is undesirable. The faces of the CF are divided by three to five borders into equal parts. Then, evaluation is performed of the capacity of scheme variation, and, if necessary, solution space is truncated.

In order to cut down solution space we unite base functions in groups. Each group is described with some characteristic function, equivalent to all the base functions of a group, which can be realized on any sub-element of this group. The base functions, realized on structurally redundant sub-elements can be included in some groups, i.e. base function $f_q = \overline{A \cdot B}$, realized on sub-element NAND, can be included in groups with characteristic functions $f_x = \overline{A \cdot B}$ and $f_x = \overline{A \cdot B \cdot C}$. Hence, groups of base functions can intersect in terms of sub-elements and isofunctions.

We provide below the generalized translation procedure for routing improvement:

1. Analysis of CF topology. Division of CF into MD.
2. Analysis of routing draft. Analysis of a subset of some non-traced connections *T*. Formation of the routing draft improvement strategy. Selection of improvement algorithm modification and optimization criteria.
3. Formation of solution space.
4. Formation of the base function group.
5. Definition of group examination priorities.
6. Group selection. Realization of reassignment operation. Evaluation and fixing of the result.
7. Assessment of the possibility and necessity of continuing, in case of termination - go to 8, else go to 2.
8. Output of the new list of connections.

For example, assume minimizing the number of tracks, passing through a boundary in a fragment of electric scheme, Fig. 3.30, a border of two neighboring MD. The functional element dimension and border between MD are defined by DMP. It can be seen that no sub-elements reallocation reduces the number of connections. The function $\overline{a}_1 \cdot \overline{a}_2$, equivalent to $\overline{a_1 + a_2}$, realized with sub-element x_{31}, can be realized with sub-element x_{41}. In other words, function $\overline{a_1 + a_2}$ can be realized with sub-element x_{41}, if at the input of x_{41} inverse outputs from x_{11} and x_{12}. can be applied. Then we have $\overline{\overline{a}_1 + \overline{a}_2} = a_1 \cdot a_2$.

In the same way we can expand the possibility of constructive realization of functions, previously realized with sub-elements x_{32}, x_{33}, x_{34}, x_{41}, x_{42}, x_{43}, x_{44}. It is obvious that solution space has not been extended. Let us reassign, using the suggested method, the functions, realized with elements x_{33}, x_{34}, x_{43}, x_{44}, x_{52}, x_{62}. Connections across MD boundary are now eliminated:

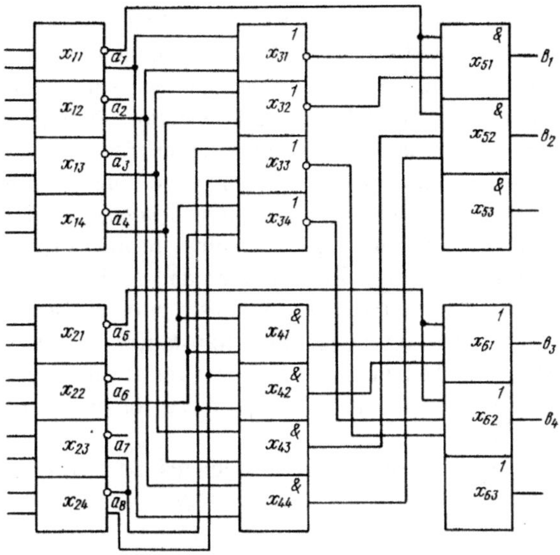

Fig. 3.30. Electric source scheme

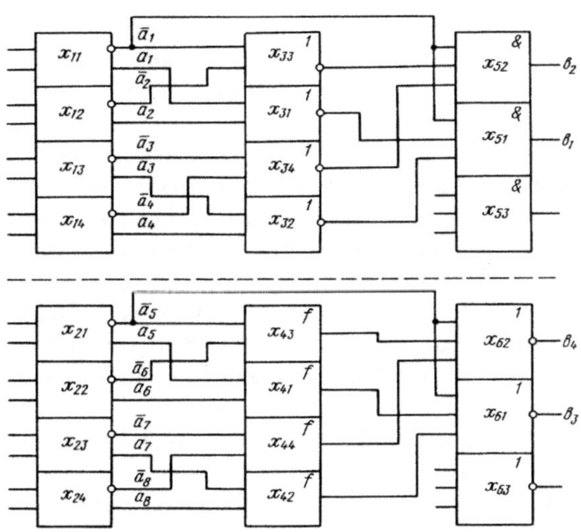

Fig. 3.31. Effects of application of the improvement algorithm

Fig. 3.31 shows a fragment of an electric scheme, equivalent to the one shown in Fig. 3.30. The algorithms, developed by the suggested methods, should be used after application of common composition, allocation and routing algorithms. In this case we secure the most effective application of algorithms, based on the method of equivalent translation.

This part contains the statement of the problem of routing of ECE elements. Classification of routing algorithms is provided, and models of commutation field assembly are described. We decompose the routing problem into four stages and describe main problems on each stage. A model of wire connection routing is developed and methods of its solution, based on Prim and Steiner trees are described. The issues of determination of planarity of the scheme and construction of a flat layout are discussed. Current methods of distribution in layers are described. Main attention is devoted to the fourth stage of routing. In this respect the flexible, ray and channel routing algorithms are examined and analyzed. At the end of this part, questions of solution improvement for construction problems of composition, allocation and routing, based on equivalent translation of logical functions, are studied. This chapter should help in an independent development of new routing algorithms and in the use of available experience for design of new ECE generations.

Connection routing is the most important of all design problems, consisting in placing definite connections between contacts without intersections.

3.7 Algorithms and Models for Design of Topology of Integrated Circuits and Microprocessors

Integration level; microprocessor; component; ESD synthesis and analysis; topological design; topology control; custom, semi-custom LSI; intelligent CAD system; knowledge base; expert system.

Nowadays, in design and production of computers, main elements are the IC. IC are microelectronic devices, carrying out conversion, signal processing or information storage, having high density of electrically connected elements and crystals, considered as a single whole. IC *element* is a part of IC, realizing a function of an arbitrary radio component (RC), realized integrally in crystal and not distinguished as independent product. The complexity of IC is given by the number of elements and components, contained in IC, and is called *integration level* (IL). The value of IL is defined as IL = lg N, where N is number of IC elements and components.

As it was said before there are six essential IC types: IC1 (first integration level): containing up to 10 elements and components; IC2: containing 10^1-10^2 elements and components inclusive; IC3: containing 10^2-10^3 elements and components; IC4: containing 10^3-10^4 elements and components; IC5: containing 10^4-10^5 elements and components; IC6: containing 10^5-10^6 elements and components. An integrated circuit, containing 500 million and more elements, produced using bipolar process, with 1000 and more elements, produced by MIS process, is called *large-scale integrated circuit* (LSI). Correspondingly, VLSI is an IC containing more than 10^6 elements and components.

Microprocessor is a software programmable device performing digital information processing and control over it, built with one or several LSI.

In production of LSI, semiconductor wafers are used. The part of a semiconductor wafer, inside and on the surface of which the IC elements, inter-element connections and contact pads are formed, is called crystal.

Use of LSI as ECE element base allows for improvement of electric and performance characteristics of ECE. In particular, their use improves processing speed and reliability; decreases power consumption, mass, cost, overall dimensions and weight. The use of VLSI improves yet these parameters and that is why VLSI are the element base of the 5th generation computers. Therefore, creation of effective LSI and VLSI CAD systems is nowadays a major problem. It is obvious that only thorough analysis and synthesis of the entire VLSI design cycle, including the entirety of work stages from exploratory design and requirements specification to production, testing, debugging and installation, will lead to creation of effective VLSI CAD systems.

The LSI and VLSI design cycle is generally divided into following stages:

1. Definition of the functional structure of the set of LSI and requirements for each LSI.
2. Simulation of logics, and synthesis of control and diagnostic tests.
3. Choice of physical structure and calculation of component parameters.
4. Synthesis and analysis of electric scheme diagrams.
5. Topological design.

Here, the set of LSI means totality of typical LSI, carrying out different functions, which are compatible in terms of architecture, embodiment, electrical parameters, and can be used together in ECE building.

In practice, LSI development is follows a familiar technological process of development, or a new technological process is created for developing the LSI set, given the required parameters.

The functional structure of the LSI set and technical requirements on each LSI are defined during the development of the ECE structure chart. A developer, using computer in dialogue mode, carries out analysis of ECE structure and defines the set of required LSI. After that, requirement specification is composed, taking into account technological capabilities, available materials and production accessories.

During simulation of logics the questions are solved of optimal or quasi-optimal selection of the entire functional basis for building different types of LSI, and for building logical structure of ECE node with logic elements of the defined basis, fulfilling required functioning rules. On the logical synthesis stage circuit realization technique, construction and technological limits are accounted for. Note that minimization of logical elements and connections between them can increase the share of non-defective chips due to decrease of crystal area. Use of high-performance logical elements basis can improve performance of the LSI developed.

LSI tests are based on functional and diagnostic analysis. To LSI input contacts test sequences of electric signals are applied. The response (reaction) of the scheme analyzed to input signals is compared with the one of non-defective (reference) LSI. During function control, workability of LSI is checked to establish whether it is defective or not. In diagnostic control certain faultiness of the analyzed LSI can be shown causing incorrect LSI operation. On LSI surface contact pads are built (control points) for intermediate control. Use of functional and diagnostic control in LSI design process allows for shortening of design time, as enabling fast detection of errors in LSI. A structural diagram of LSI design process, including chip control, is shown in Fig. 3.32.

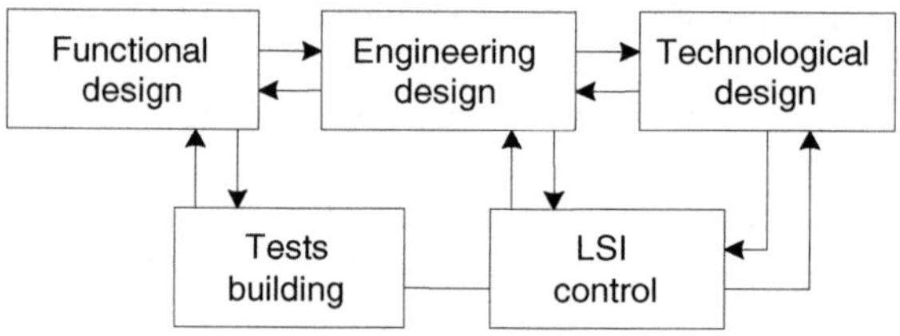

Fig. 3.32. Structural diagram of LSI design process (including control)

Simulation of logics on element (most often, gate) level is done in LSI analysis to check the realization of required logical functions taking into account time delay of signals.

Let us examine in brief the problems of choice of LSI physical structure and calculation of component parameters. The physical structure of LSI includes diffusion profile, electro-physical parameters in different diffusion layers, etc. This structure is defined with respect to the requirements as to the main transistor, included in LSI. Then, for calculation of other components (other transistors) the selected structure of the main transistor is used. It is considered that changing only of component geometry is a simple method (ACC≈$O(\alpha_n)$) of obtaining different component characteristics. Note that for practically all LSI in a set, use of separate component geometry is tried out. In the future, it will be required in LSI design to have values of electric parameters that describe models of components, used for synthesis and analysis of electric schemes.

In some cases, to obtain samples of active components, for which parameters are measured, special topological test schemes are developed. They contain active and test components to control physical structure parameters and possible production faults. On their basis different types of geometric configurations are selected that fit schematic requirements the best, statistical data are obtained on parameters of active components in different operation modes, parasitic connections between LSI components are analyzed, control photomaps of production process obtained, etc.

In this context, the problem is to develop high-speed algorithms for calculation of parameters of different geometric configurations of active components.

Let us examine the problems of synthesis and analysis of electric scheme diagrams (ESD).

First, a functionally complete series of logical elements is defined and, depending on combination of parameter values meeting LSI, logic type is selected. This logic type defines the ESD of logical elements. Then, component parameters values are found, meeting requirements on LSI parameters. Solution to this problem is based on boundary test method. For this purpose, the ratios connecting scheme parameters with those of components are determined, and the area of scheme well-behavior is defined with required parameters. The equations, connecting the scheme parameters with the component parameters, are established to carry out the analysis of the scheme:

$$y = p\ (a_1, a_2 \ldots a_m), \tag{3.9}$$

where a_1,a_2...,a_m are component parameters defining the state of the scheme, and p is an operator. In general case, solution (3.9) is obtained as a solution of a system of non-linear differential equations.

Nowadays, statistical methods are widely used in LSI analysis. In digital LSI control the standards for measuring parameters and their boundary values are set such that provide maximum share of non-defective LSI output in terms of functional parameters.

The problem of LSI ESD synthesis consists in defining electric configurations of the optimal scheme. This is obtained from RS, with LSI defined by the function realized and the requirements as to its parameters.

The system of LSI analysis should have a simple language for describing ESD configurations and types of components, a large MM set, the ability to change automatically parameters for carrying out sensitivity and tolerance analyses, etc.

Topological design of LSI consists of three main stages:

1. LSI topology synthesis;
2. Topology control;
3. Development of control information for technical documentation production.

The source data for LSI, VLSI and MP topology design (synthesis) is constituted by ESD, geometric parameters of active components, nominal values of resistors, case type, contact pad output sequence, construction and technological limitations (CTL). Here, the problems are solved of LSI component topology design, division of ESD into parts, design of each ESD part, and topology design of the whole ESD. Component topology is selected on the basis of electric characteristics. A developer uses design solution files, database and library of typical component topologies.

Topology control includes maintenance of CTL accuracy (allowed sizes of topological shapes, overlaps, gaps, etc.), one-to-one mapping of obtained topology and source ESD, and calculation of heterogeneous influence of parasitic elements on electric parameter quality. Afterwards, development of photomaps and LSI production take place.

There are different LSI types as to their purpose and the design methods:

1. Custom commercial (produced in very big series). The design is at the instrument level. It is specific in that every element, its allocation and connections between elements, are constructed. In this case the highest integration level is achieved. This problem requires a lot of effort to solve.
2. Semi-custom or cellular type. In this case there is a set of topology fragments (cells), the topology of counter, adder and trigger. It means that there is a set of photomaps that can be used to reproduce scheme fragments on LSI crystal.

The LSI design problem is to cover electric scheme with cells from available sets, to allocate cells on crystal and to trace connections between cells. This method involves lower level of effort and is easier to automate, but the LSI integration level is also lower than for commercial LSI.

3. Founded on base crystal. Base crystal is a blank with already realized semiconductor structures, all of the same type elements. For the user, the problem is in design of connections between these elements to make LSI perform the required functions.

Because of increasing complexity of VLSI and LSI, their construction and design methods increasingly meet the requirements of automated design.

Another modification of the base crystal type is array LSI, with elements that are groups of unconnected components (transistors). During design of array LSI not only connections between elements are developed, but also connections between components inside elements. The peculiarity of array LSI is that they consist of a high number of the same width and height (or same height), same type elements, arranged in commutation field with a regular structure of slots, simplifying the processes of formalization, development and realization of allocation and routing algorithms.

The problem of LSI design for base crystal, containing array of the same type topological fragments, consists of two stages. The first stage is base crystal design, assuming comprehensive solution to the following issues:

- rational selection of functional contents of cells;
- specification of the cell number, and of buffer coordination circuits and their allocation on crystal;
- topological allocation of cells and methods of adjusting them;
- specification of power line allocation, the step of routing of interconnections and their allocation;
- determination of the required number of external outputs, selection of crystal assembly method, etc.

The efficiency of use of array LSI is largely defined by the capabilities of cells (functional elements).

The optimal set of functional elements, providing for realization of any circuit of a given device, is based on the set of logical schemes, characterizing the class of certain type of devices, and also list of requirements and limitations. As a result of the first stage, metal patterns of functional elements, based on cell components, are obtained, and performance of the required functions by each element is secured. The variants of metal patterns of functional elements form a library, introduced into the CAD system database.

In the second stage, the problem of covering of the functional scheme of LSI with the elements from the library is solved first. Most of the covering algorithms are based on selecting, from given functional scheme, the sub-schemes (groups of logical elements, maximally connected with each other), analysis of all or sufficiently many of them, and check of coincidence of logical functions and logic elements from the set used among the sub-schemes. A sub-scheme is assigned to the functional element that realizes most of its logical functions. The process is repeated until the elements of functional scheme are fully covered by functional elements. The problem of covering of the functional scheme of array LSI does not differ from the one of covering of cellular LSI functional scheme or of covering the functional schemes of equipment on PCB with chips. Then, the problems are solved of topological pattern allocation on crystal and of design of connections, accounting for all construction and technological limits. The second stage, by introducing changes into the metal layer patterns, defines the function of each LSI.

From the algorithmic point of view, there is continuity between this problem and the one of equipment design, based on PCB with microchips. However, significant increase of problem dimension, connected with sharp increase of the number of allocated elements and the size of commutation field, as well as possible irregularity of external output allocation of each element, require a revision of algorithms and software, developed to date.

The essential problems, solved by modern automated design systems for LSI based on base crystal, are:

1. Preparation and input of the library of elements used;
2. Control and correction of source information;
3. Translation of source information description into internal format and saving it in an archive;
4. Topological design of inter-circuit connections (allocation, routing);
5. Full control of inter-circuit connections topology;
6. Translation of description of topology to a certain production system language, generating photomaps;
7. Interaction of designer with computer during design.

The primary difference between CAD systems based on crystals and CAD systems for LSI with elementary circuit components consists in wide use of allocation and routing programs for designing interconnection topology and in a significant decrease of photomap design labor input.

The most important factor that defines efficiency and viability of a LSI automated design system is the presence of highly flexible, in terms of program and technical aspects, control subsystem for design results. Programs checking continuity of conductors and gaps between conductors establish, whether requirements are fulfilled on width of conductors and distances between them, and do connections, realized in metal patterns, correspond to those from the logical structure. The assembly requirements control program examines each circuit for loads, connection lengths and metallization width, and forms the list of circuits, that do not meet the requirements. Crystal designer revises circuits with disturbed parameters.

From the mathematical point of view, composition, allocation and routing problems are NP-hard combinatorial problems. The essential problem encountered in LSI design is high computational complexity of design problems, linked with large dimensions of the object designed. One of methods of reducing complexity is decomposition. The decomposition can be based on circuits of the scheme or on problem kind. In decomposition by the kind of problem solved, the entire problem is divided into several different type problems, each having its own specific features. Solution to each problem is obtained by its local criteria, which, however, should not be in conflict with global optimization criteria.

In scheme-based decomposition, the complex optimization problem is divided into similar problems, solved for separate sub-circuits. Thus, in design of LSI topology, division into nodes that carry out independently complete functions is done, and their topology is designed separately. Then, the solutions obtained are connected. The size of sub-circuits is defined by the capacity of the basis optimization procedure for solving definite kind of problem (e.g. allocation, routing). For this purpose, the method of optimal reduction can be used, in which a source problem of large dimensions is divided into hierarchically embedded identical problems of significantly lower complexity.

The third manner of reducing problem dimensions consists in shrinking the set of initial solutions by shedding unpromising variants. Thus, during routing problem solution for each connection a set of realization variants is generated, significantly differing as to the commutation field resource distribution. Other variants, whose number can be much greater than of those in the set generated, are excluded from further examination. Search for routing solution is performed in the generated set of connection variants. Most of the solution methods for LSI allocation and routing problems, developed to date, are based on these principles.

An intelligent CAD system (ICAD) is a technically organized system that consists of a set of intelligent design automation tools, linked with design organization subdivisions, and performing automated or automatic design. It consists, therefore, of a set of tools and components of technical, mathematical, software, information, and language, methodological and organizational nature. An intelligent CAD system should carry out automated or automatic design in all respective stages, based on use of mathematical models, design procedures, computer and organization techniques. In order to enhance the "intelligence" of a CAD system, extend its capabilities and improve interaction with the user, it is necessary to fill it with domain knowledge of design and application. It is also desirable to supply the CAD system with tools for conversion and practical use of introduced knowledge as well as with methods for summarizing, learning and self-learning. One of the most important CAD components should be the system configuration unit for optimal design (in terms of predefined criteria) and evolution ability concerning given requirements.

A general structure of ICAD, like structures of other intelligent systems, consists of three subsystems: interaction, processing, control. The interaction subsystem uses input description analysis together with available knowledge and forms internal incomplete problem representation. The processing subsystem transforms incomplete description into a complete one using interaction with the control subsystem and transfers it again to interaction subsystem. Then, the process iterations continue until satisfactory solution is obtained.

Modern ICAD systems should be configured for the considered problem and technological process, with the ability of flexible reconfiguration to other objects and new technologies of VLSI manufacturing. The structural chart of ICAD is shown in Fig. 3.33. The system can be used by a number of designers and technologists. The dialog processor (DP) is a structural unit of intelligent input-output

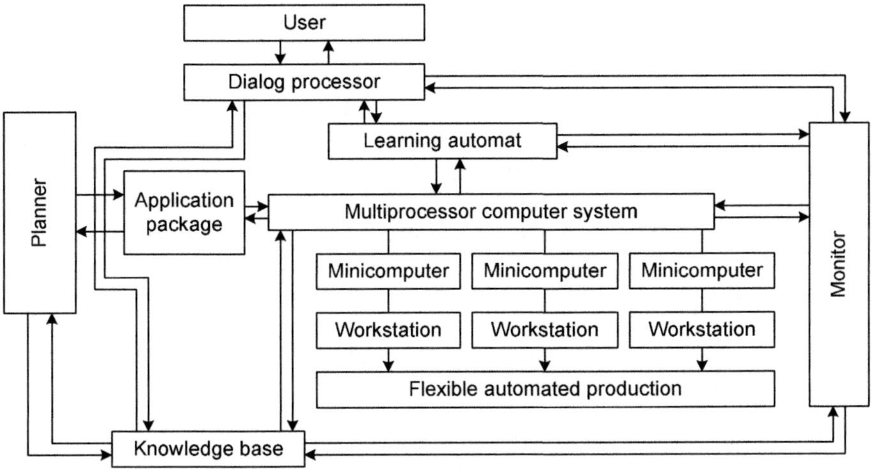

Fig. 3.33. Architecture of an intelligent VLSI CAD system

interface. It provides the required dialog script during design. Dialog forms differ depending on user's knowledge level. For this purpose, the database contains different hierarchically organized supplementary information. The dialog processor should provide speech and visual input-output. The best form of DP realization for VLSI design is combined software plus hardware implementation. Dialog processor should have a set of dialog scripts, hardware generator of random and directional search, and enumeration processor. A user, interacting with DP, divides the VLSI design problem into functionally different sub-problems. Solving of local problems with optimization of selected local criteria is carried out in series or simultaneously. The main purpose of the planner is information search. Together with knowledge base (KB) the planner carries out search procedures to find solutions and to select the best result. The planner uses expert system to analyze DP source data. Depending on equipment class, design parameters, defined work modes, etc., a solution path is selected. Each solution path corresponds to a control program unit in the monitor. This unit controls selection, editing, configuration and connection of different software units, definition of types and formats of intermediate data, and formation of the working program.

An intelligent CAD system contains not only KB, but also DB. This is caused by the necessity of processing and storing large volumes of formal and numerical reference data, invariant with respect to production technology. During ICAD functioning, working data set (WDS) is formed, containing information from KB and DB, required on the current stage. The computer network is a set of equipment, including supercomputers, minicomputers, and local area network of personal computers; their operation mode during CAD operation is defined by the planner. The results of design, after control, verification and editing are entered into LSI ICAD design solutions archive and arrive at the CAD output as technical documentation and control documents on machine data carriers.

KB is the central ICAD component. Its main purpose is to acquire and use knowledge, introduced before the design process, obtained during the design process and representing additional knowledge for the CAD and the user. KB is a set of certain facts and information pieces united into one software and hardware environment. The facts and information in KB define certain knowledge domain (in our case, VLSI). It is considered that KB consists of three main elements: rules, queries and transformations. There are three kinds of KB, the bases of generic, system and application knowledge. The first ones store basic knowledge required for solution of all design problems, the second - knowledge of the system, the third - all application-specific knowledge, for example, enterprise description, design rules, methods of technological design, certain algorithms, their typical blocks, etc. Unlike standard databases, KB should be flexible, besides, it should allow for processing information and forming new knowledge. Knowledge base can be seen as a chain of frames. The first frame unites all knowledge about current design situation and predicts which objects will be processed and what events can occur. In our case this frame is the whole of the procedure and declarative knowledge about VLSI models and will consist of many slots, i.e. frame information structures. The slots contain knowledge about certain VLSI design situations; perform application calls from intellectual CAD (ICAD), together with monitor control information input and output. Slots of one frame can be connected with slots of other frames, forming graph (binary connections) or hyper-graph (n-ary connections). Optimal search and conversion of information in such graphs requires selecting families of independent, dominating, isomorphic and complete subsets.

In VLSI CAD the designer model can be seen as ordered 3-tuple Mk = $<Z, L, W>$, where L is a set of linguistic variables describing VLSI design knowledge domain; Z is knowledge, defined as ordered 2-tuples $Z = \{<s_i, p_i> \dots\}$, $i \in I = \{1, 2 \dots\}$; s_i is a reference situation; p_i - a solution; W is the set of output rules. On the basis of investigation of set $S = \{s_1, s_2 \dots s_l\}$ of reference situations certain realization of device designed is selected. The set $P = \{p_1, p_2 \dots p_l\}$ defines the rules of passing from situation s_i to s_j, W allows for providing well-founded output from solution $p_i \in P$ in real-life situation of a design process. Note that Mk is based on ES enabling a combined scheme of all the problem solution processes. Such design technology allows for composing of a small set of leading questions that will help the user (designer) to pass through all the design stages.

The KB hardware support allows for maximum performance in realization of complex design algorithms. Two problems are most important in this context. The first one is search for and development of new algorithms, adapted the best to hardware realization. The second one is physical realization of algorithms. The issue of establishing the adequate proportion between hardware and software parts of ICAD naturally arises.

Development and investigation of ICAD systems resulted in application of artificial intelligence systems using knowledge from local knowledge domain for solution of problems that occur in it. Such systems are called expert systems (ES). Nowadays, ES are used for solution of interpretation, diagnostics, prediction, planning, management and design problems. They can already now consult and give advice, instruct and learn, carry out search, analyze and classify, represent information in a required way, build projects, forecasts results of ICAD work, produce new design concepts and technologies. The essence of ES is hosting of heuristic, formal and rough knowledge coming from specialists in a domain of knowledge, and using this knowledge, for example, during ICAD problem solution.

In a general view, ES consists of four components: KB, output machines, knowledge extraction unit, explanation system (interface).

For determination of logical output ES use straight reasoning chains from data to hypothesis and reverse reasoning chains, with data required to prove or disprove some hypothesis. Sometimes combinations of these two are used, e.g., indirect reasoning chain. The essential participants in the ES operation are: an expert or a group of experts, users, verifier (i.e. software or hardware device providing feedback between expert and ICAD), engineer-interpreter. The usual requirements as to VLSI ICAD ES are:

- ability to reason with uncertain data;
- limitation by certain examination environment;
- ability to give clear explanations;
- extensibility;
- presence of clear advice at the output;
- performing of expert questioning based on method of questioning and estimation of completeness and consistency of information obtained;
- filling of ES KB with acquired information;
- establishing the explanation and planning system configuration for working with problem domains.

A VLSI ICAD expert system is a whole of interacting local and autonomous subsystems, each of them describing the corresponding sub-domain of the problem domain.

In conclusion we note that integration of hardware and software instruments in ICAD allows for increase of design efficiency and speed.

3.8 Topological and Metric-Topological Approaches to LSI Design

Linear, fan-shaped, star-shaped circuit configurations; transparent vertex; planarization; analysis of graph planarity; chip area.

Design of topology of LSI and VLSI is the process of transformation of electric circuit information into level-proper geometrical information that specifies allocation and connections of components of a scheme on a chip.

Traditionally, LSI design process was divided into execution stages, i.e. problem decomposition was performed, with paralleling of data processing and decision functions between separate elements of the design system.

The topological approach includes following main problems that split the design process into a sequence of stages:

- construction of the mathematical model of the scheme;
- analysis of planarity of the scheme model;
- formation of the flat layout of the scheme model;
- planarization of the layout;
- synthesis of geometry of the scheme.

In the initial stage, for the LSI circuit diagram a mathematical model is developed, reflecting topological properties of the scheme. The apparatus of graph theory is mainly used. The primary issue in modeling is that a model be adequate to its object. The quality of the entire design process depends on the completeness of reflection of properties. That is why the questions of model building are fundamental in topological approach.

In this context, the main criterion of model selection will be description adequacy, meaning that planarity of the model is the necessary and sufficient condition for planarity of the scheme. We shall say that a scheme model is planar if there exists its flat topological representation. Such definition makes it possible to:

- connect model planarity with existence of at least one topological presentation of it;
- pass in a unique manner from planar topological presentation of the model to planar realization of the scheme.

The generally accepted approach to the analysis of topological properties of the scheme includes decomposition principle, consisting in separate examination of models of scheme components. LSI components are, in general, elements with developed topology, electric connections and external contact pads.

For the functional elements the simplest model can be constructed by mapping the elements as graph vertices.

A more accurate element model is built by mapping the functional element contacts as vertices X of a graph G. In the element model the set V of edges is introduced to represent one-to-one belonging of contacts to this element.

Such a model allows for:

- considering different topological descriptions of cyclic, linear and combined element sets;
- defining strict contact order in an element;
- representing equipotential contacts and contact groups, for which the sequencing in the element routing cycle is characterized by lack of crossings ratio; equipotential contacts are formed by group of outputs, electrically united inside it;
- representing invariance of adjacent contacts and their groups; invariant contacts are a group of logically equivalent outputs;
- defining equivalence and equipotentiality of outputs and output groups.

Electric connection in LSI is a loop uniting respective contacts of elements. Main requirement for connection representation is integrity of connection contacts. If connection is interpreted by a Steiner tree, the overall number of connection trees increases significantly. That is why arbitrary selection of one topological variant of connection during its model mapping, leads to representation ambiguity. The model should allow for consideration, during planarization, of all admissible configurations of the connections tree.

An adequate description of multi-contact circuit can be given by a hypergraph edge representing connection as set of contacts, belonging to it. However, during analysis of model planarity or construction of flat layout, multiple interpretations of connections cannot occur. We must have planar topological representation of connections. Three configurations of trees, representing electric connection, are distinguished in the analysis of planarity of electronic circuits: linear, fan-shaped and star-shaped (Fig. 3.34).

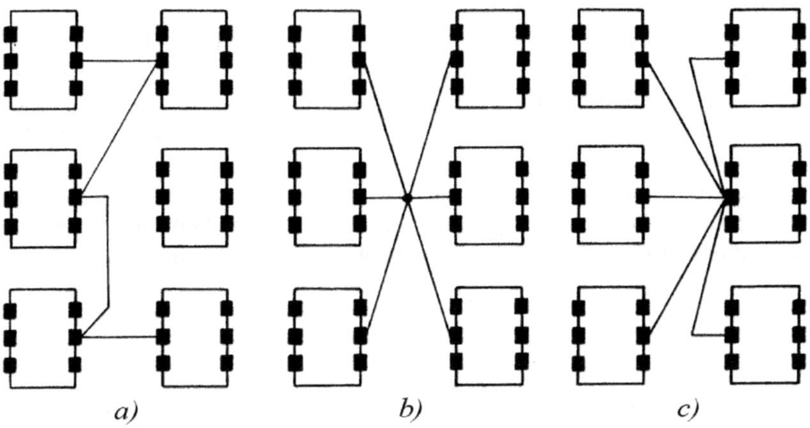

Fig. 3.34. Different circuit configurations: a — linear, b — star-shaped, c — fan-shaped

Absence of branching vertices and sequential vertex connectedness are typical for the linear tree. Such circuit model allows for representing only one possible configuration of its topology.

The star- and fan-shaped tree configurations are formed by connecting all vertices with the central one. For fan-shaped tree any vertex that represents circuit contact can be selected as the central point. For star-shaped tree an extra point should be introduced.

The disadvantages of the star-shaped representation result from the presence of the "transparent" vertices in the model of scheme components. A vertex is transparent if in a sufficiently small neighborhood there are more than one connected domains for laying out edges representing circuits. The use of the "fan-shaped" circuit representation with the center in a transparent vertex allows for eliminating this disadvantage, if the set of contacts incident to the circuit contains at most one transparent vertex.

The stage of analyzing the planarity of the scheme model is carried out for flat representation of the scheme. This problem is senseless for schemes modeled with non-planar graphs. In such a case the problem is to determine largest set of non-intersecting connections, i.e. the goals of the second and third stages of topological approach to design are associated.

The problem statement for determining the largest set of non-intersecting connections is formulated in the following way:

A graph G = (X, U) and a positive integer $k \leq |U|$ are given. We want to find a planar subgraph G' = (X, U') in the set of all possible planar subgraphs, for which $k' \leq |U'|$. This problem is NP-complete, but when $k = |U|$, solution can be found in polynomial time, i.e. the problem of checking planarity of a graph belongs to the class of polynomial problems. The mathematical apparatus for checking graph planarity is well elaborated. Therefore, for finding the maximum flat part of the VLSI scheme the formalized procedures of graphs planarity analysis are commonly used.

The methods of graph planarity analysis belong to two classes. The first class includes methods based on criteria of planarity providing necessary and sufficient conditions of graph planarity of Pontryagin-Kuratovski, McLain, Whitney, Harary-Tatt. These methods are of little use from the practical point of view because of high computational costs.

The methods of the second class are more effective. They are based on the algorithmic search for the solution of graph planarization problem and allow for performing the planarity analysis of a given graph step by step. Three methods are distinguished:

1. The method by L. Auslander and S. Patter and its modifications, suggested by I. Hopcroft and R. Taryan. This method is based on the search for a cycle in the graph. After removal of the cycle, the graph is divided into disconnected parts (segments). Each segment, related to the inner or outer border of the selected cycle is analyzed for planarity. To enhance performance, the strategy of single-direction branching is used. A modification of this approach is described in Section 3.3.
2. The method by Demukron, Malgrange and others and its realization developed by Rubin. This method analyzes planarity through sequential embedding of external components (separately arranged parts of graph) into individual faces of the flat representation.
3. The method of Lempel, Evan and Cederbaum, with layout construction beginning from some vertex, with all its incident edges connected to it, then, the vertices, incident to these edges, and so on, until layout of the whole graph is obtained. Reduction of edges to cycles is done relatively to infinite face of layout being already built.

The main advantages of the here reviewed methods of analyzing graph planarity are:

- orientation at investigation of graphs without limits;
- guarantee of correct check of graph planarity.

The stage of planarization consists in realization of connections removed during flat layout building. Planarization of the non-planar edges is done using production techniques:

- by introduction of extra elements ("diving" elements, bridges, etc.);
- by building connections over (under) elements;
- by invariance of contact groups of different elements of the designed scheme.

The source information for the planarization problem is constituted by the graph G of the scheme, defined by flat layout G_i^* and set of removed edges U_{rem}, and construction-technological methods of eliminating crossings, defined by the technology used.

The planarization problem is formulated as follows in graph terms:

Planarization of a non-planar graph G = (X, U) is an operation or a sequence of operations F = {$\varphi_1 \ldots \varphi_k$}, as a result of which G can be completely represented by the flat layout G^*, i.e. $G \xrightarrow{\varphi_i} G^*$.

There are following types of graph planarization operations:

introduction of extra vertices: $G = (X,U) \xrightarrow{\varphi_1} G(X^*,U)$, where $X \subset X^*$;

introduction of extra edges: $G = (X,U) \xrightarrow{\varphi_2} G(X,U^*)$, where $U \subset U^*$;

introduction of extra edges and vertices: $G = (X,U) \xrightarrow{\varphi_3} G(X^*,U^*)$, where $X \subset X^*$ and $U \subset U^*$;

graph structure change: $G = (X,U) \xrightarrow{\varphi_i} G^*(X^*,U^*)$, where $X \equiv X^*$ and $|U| = |U^*|$.

Let us examine the use of operation of introducing extra vertices.

We have: G' = (X', U), a flat subgraph of graph G = (X, U), and a set of non-planar edges U_{rem} on G; $\forall \alpha_i \in \Lambda$ the subgraph can be converted to flat graph Gi = (Xi, Up), so that $\forall x_j \in$ Xi \ X is associated with connected subset $K_i^E \leq K^Q$ of the intersection set KE $\Rightarrow U_{rem} \cap U' \neq \varnothing$. Find the layout variant $\alpha_* \in \Lambda$, for which |Xi \ X| → min.

The existing methods of solving this problem include the following procedures:

1. Building of spatial model for construction of edges.
2. Rating of the list of edges not included in flat layout.
3. Finding the shortest path between the vertices of the constructed edge.
4. Modification of space after each edge is constructed.

We use as spatial model for construction of edges the flat layout, corresponding to a solution, represented by the set of faces, which contain no edges. There are two approaches to conversion of the spatial model:

- model elements are replaced by edges and extra vertices - by spaces, the topological working field is obtained;
- a metric is introduced on the set of edges, distance between adjacent edges is taken as a unit, a discrete topological working field is obtained.

In topological space thus defined the planarization problem turns into the one of finding the shortest paths. A path between two vertices in the working field is defined as a repetition-free sequence of neighboring vertices or faces. The search and shortest path building procedures are similar to the corresponding procedures of connection routing methods. After each connection is constructed the working field must be corrected. The process of working field modification can end with replacing each face, through which connections are built, by two new ones, or introduction of new vertices and division of the field along the path obtained. Obviously, the total number of intersections of the edges constructed will depend on the order of their construction.

In the final stage of the topological approach the problem is solved of constructing the geometric outline of the scheme topology. Here are some indications as to the way this problem is solved:

1. The image of the flat layout, obtained in previous stage, is automatically formed. This allows the designer for building geometrical outline of topology in an interactive mode.
2. Geometrical outline is formed automatically based on the assumption, that a planar graph, independently of allocation of vertices, has always a flat layout. The problem of realization of connections for a definite allocation is considered here. The disadvantages of this method are: extension of connection tracks, excess of intervals between elements and connections; and ultimately even the necessity of rearranging the elements because of impossibility of routing.
3. Arrangement of elements and routing of connections are done with respect to features of flat layout. For example, the allocation problem is solved first. Input data are constituted by the flat graph of scheme connections and information on relative distribution of elements. Elements are compactly allocated in a rectangle. The size of an element is the area, required by this element, expressed in MD. The MD size is defined by element set of the scheme realized. MD are allocated in rows and columns of the rectangle. A model is built on the basis of initial results of allocation and topological routing is performed in it. Iterative process takes place: construction of a routing model and preliminary routing in it. Final variant of connections on the plane is obtained by considering model routing results after an acceptable variant is found.
4. The geometrical outline of the known flat layout can be obtained using metric-topological design approach as follows: Let us have in some design stage a set of allocated elements, and an orthogonal border on chip assembly field characterize allocated elements and their connections, with all non-branched connections belonging to the current border. To minimize chip area and overall connection length, it is advised to place the next element in routing area as close as possible to the current border. To do this, a corner point is found on the border with minimal coordinates. An element is taken as the next one if it was not allocated and tested in all allocation directions. After an element is preliminarily fixed, the possibility of its placing and its exact location are checked. New current border envelopes the placed element, and its non-traced circuits belong to border part. These steps are repeated until all elements and connections are allocated; AAC is within O (n^2)-O (n^3).

3.9 Synthesis of the Metric-Topological Model for Array LSI Topological Outline

Metric-topological model; topology fragment

Topology outline (TO) design is the dominant problem of LSI design, as it defines the possibility to design construction realization of the scheme, quality and manufacturability of following production. Scheme irregularity and increasing tendency towards component integration determine high dimensionality and complexity of this problem.

One of possible approaches to realization of topology and connection structure is the cellular approach, allowing for introduction of element irregularity in topology organization and for reduction of the problem to the one of allocation and synthesis of typical topology fragments (FT). The main difficulty in automated design of array LSI is the necessity of accounting for a high number of factors and limitations.

These factors and limitations make practically useless the methods of allocation and routing on discrete working field with strict fixation of elements and connection fragments, and require a new approach. Association of solutions to problems considered requires development of one, simplified and efficient routing model that would allow for dividing the intricate LSI topology design problem into smaller and simpler sub-problems, with solutions ensuring fulfillment of the common design goal.

One variant of automated design of LSI topology is composed of three main stages: development of the common model of LSI FT allocation and connections; optimization of LSI connections topology; and combined LSI topology striping. Let us examine the first stage. It suggests solving the problem of array LSI FT allocation and connections by the synthesis of a common metric-topological model of topological field. The main concept of the approach is to create favorable conditions for the second stage, optimization of topology of connections.

The input to allocation is constituted by the connections diagram of FT, corresponding to elements of the functional scheme of the device designed. We have a set $X = \{e_i \mid i = 1, N\}$ of FT, differing as to type $t_\gamma \subseteq T$, $|T| = w$. This set is in general case divided into w subsets of same type elements $X'_1, X'_2 \ldots X'_\gamma \ldots X'_w$; $|X_\gamma| = \beta_\gamma$. Select two subsets: $X_0 \subset X$ — elements of initial allocation, and $X_1 \subseteq X$ — elements to be allocated on free chip area, $|X_0| << |X_1|$. An element $X_i \in X'_\gamma$ is characterized by height h_γ and width b_γ, with $b_\gamma = b'_\gamma + \Delta b_\gamma$, $h_\gamma = h'_\gamma + \Delta h_\gamma$, where b'_γ and h'_γ are actual FT sizes, Δb_γ and Δh_γ are reserves for routing (experimentally defined). In cellular design h'_γ is averaged to

$$h = \max_{t_\gamma \in T} h_\gamma .$$

To assess the quality of allocation a combined criterion can be introduced:

$$K(M) = \sum_{i=1}^{\xi} \mu_i K_i(M) , \tag{3.10}$$

where K_i (M) are partial criteria of allocation quality; M is selected model of designed structure; μ_i are weights of criteria.

The problem of FT allocation consists in finding such arrangement of the elements of set X on a plane that the overall length of connections is minimal. To obtain a correct problem statement it is advised to estimate the area and overall dimensions of the LSI crystal necessary for allocating all elements of the set X. These requirements are the main constraints on solution. A conclusion can be made on the perspective of combination of initial constructive allocation with following iterative optimization. During realization of such approach, the array LSI FT allocation problem can be represented by a sequence of sub-problems: calculation of required area and overall dimensions of LSI crystal, constructive allocation of FT on the crystal and its iterative optimization.

Design of a LSI crystal is characterized by working field area $A_F = H_F B_F$, where H_F and B_F are, respectively, height and width of the field. The number of horizontal FT rows (lines) is defined by $K_F = \left] \frac{H_F}{h} \right[$.

The elements of X_0 are divided, in general case, by lines numbered k ($k = 1, 2 \ldots k_F$) into η_k allocation zones of width $B_{k1}, B_{k2} \ldots B_{k\eta_k}$, forming actual line width:

$$B_k^A = \sum_{\eta=1}^{\eta k} B_{k\eta} . \tag{3.11}$$

The FT allocation problem is solved with two models in rectangular coordinates. The first model is the field Q, limited by coordinates $S_{\min} = t_{\min} = 1$, $S_{\max} = t_{\max} = \left\lfloor \sqrt{N} \right\rfloor + 1$, where points define conditional allocation of X_0 elements; and distribution by crystal side by points with one coordinate is undefined and is conditionally considered to be zero. The second model is a rectangle R with dimensions H_P^T (vertical) and B_P^T (horizontal). These are typical sizes of chip working field, where, considering overall dimensions, X_0 elements are allocated (Fig. 3.35). Parameters $B_{k\eta}$ are assigned to the established allocation zones.

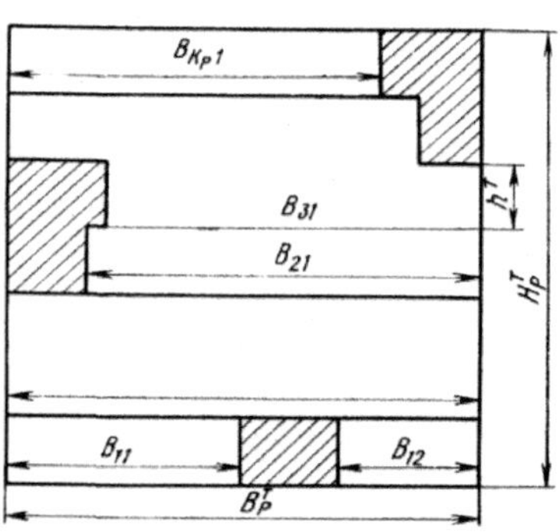

Fig. 3.35. FT allocation model accounting for overall dimensions

In calculation of chip area and overall dimensions, first the expected chip area A_F is defined: the areas of elements X_0 and X_1 are summed up and

$$\sigma = h_t \max_{\gamma \in T} b_\gamma \sum_{k=1}^{k_F} \eta_k \,. \tag{3.12}$$

Then, with k_F and h, the sizes of constant chip part, and considering the technological discrete d of chip dimensions, the overall H^T and typical H_P^T, h^T values are determined. A_F and d, given the dimensions of the constant part, allow for defining the size B^T, and then — size B_P^T.

Constructive FT allocation is done in two stages: preliminary allocation of parts corresponding to elements $X_i \in X_1$ on field Q; and precise determination of allocation of elements X_1 on model lines.

To arrange with the part model we use an approach based on a mechanical analogy. Points $e_i \in E_1$ are selected sequentially in accordance to their connectivity with other elements. The data come from adjacency matrix $\mathbf{R}_E = \left\| r_{pq} \right\|_{N \times (N+\varphi)}$ of elements X, where φ is the number of external outputs from the scheme.

Each point x_j obtains allocation coordinates, calculated by formulas, similar to known formulas for the center of mass of a particle system, where, instead of mass, the generalized weight is used of x_j contiguity with allocated elements and scheme EO:

$$s_j(t_j) = \frac{E_{s(t)} + \sum_{i=N+1}^{i=N+\varphi} s_i(t_i) r_{ij}}{\sum_{\forall e_i \in E_0'} r_{ij} + \sum_{i=N+\varphi}^{i=N+1} r_{ij} u(s_i(t_i))} \,, \tag{3.13}$$

where

$$E_{s(t)} = \sum_{\forall e_i \in E_0'} s_i(t_i) r_{ij} \,; \tag{3.14}$$

$$u = \begin{cases} 1, \text{if } s_i(t_i) \neq 0; \\ 0 \text{ otherwise.} \end{cases}$$

This concept is implemented by an algorithm with the following logical scheme:

$$A_0 \downarrow^2 A_1 A_2 w_1 \uparrow^1 A_4 w_2 \uparrow^2 A_k .$$

Here A_0 and A_k are operators of algorithm start and end; A_1 is selection operator of element x_j from matrix **R**, such that

$$\forall_{x_i} \in X_1[e_j \Rightarrow r_{jj} = \max_{1\le i\le N} r_{ii}]\ ;$$

A_2 is an operator for calculation of preliminary allocation coordinates using formula (3.13); A_3 - assignment operator, $s_j := ½(S_{max}+1)$, $t_j := ½(t_{max}+1)$; A_4 - operator for forming subsets $X'_0 = X_0 \cup e_j$, $X'_1 = X'_1 \backslash X_j$, and assignment $r_{jj} := 0$; logical conditions $w_j = 0$ corresponds to situations $w_1 - |E_s| = |E_t| = 0$ and $w_2 - |X'_1| = 0$; when $w_{jj} = 1$ - jump by pointer.

As the result of FT allocation we obtain description of relative allocation of elements X_1 in model lines allocation zones.

During iterative optimization of FT allocation on the model, additional information concerns the width of LF allocation zones. The target function of allocation is $F = \sum_{q=1}^{q=n_c} \tilde{L}_q$, where $\tilde{L}_q$ refers to length of circuit q; n_c is the number of circuits in the scheme.

Conditions of element pair rearrangement are as follows:

1. The (i,j)-rearrangement of elements X_i and X_j is fixed if $\Delta F(i,j) = F_{cur}-F(i,j) > 0$, where $\Delta F\ (i, j)$ is function change for (i, j)-rearrangement, F_{cur} is current F value; $F(i,j)$ - value of F after (i, j)-rearrangement, given the limitations on FT and LF width.
2. If $\Delta F\ (i, j) = 0$ and limitations are fulfilled, (i, j)-rearrangement is fixed if it decreases width difference of LF allocation zones, i.e.

$$\Delta B\ (i, j) = |\Delta B_i\ (x_i) - (\Delta B_j\ (x_j)| - |\Delta B_i\ (x_j) - (\Delta B_j\ (x_i)| > 0, \tag{3.15}$$

where $\Delta B(i,j)$ is the decrease of LF zone width difference after (i, j)-rearrangement; $\Delta B_i\ (x_j)$ is LF zone width after moving element X_i to place of element X_j.

Condition 2, in general case, accompanies Condition 1 during consecutive rearrangements.

To set up pair rearrangements process, arrange the FT of neighboring M_T model cells as a tuple $\lambda = \langle x_1, x_2 \ldots x_i \ldots x_n \rangle$, where i is positional number of the element (an element has also a unique schematic number). During fixing of rearrangements position numbers of elements are interchanged.

Assume that cycle i is a rearrangement sequence of element x_i with other elements, x is the main, and an arbitrary element x_j - a complementary element of pair rearrangement. In cycle i we will use only elements x_j when $j > i$ as complimentary.

The iterative allocation optimization algorithm can be represented by the logical scheme:

$$A_0 A_1 \downarrow^1 A_2 \downarrow^2 A_3 w_1 \uparrow^3 A_4 w_2 \uparrow^4 A_5 w_4 \downarrow^6 \downarrow_5 A_6 \downarrow^3 \downarrow^4 \downarrow^6 w_5 \downarrow^2 w_6 \downarrow^1 A_k,$$

where A_0, A_k are operators of algorithm start and end; A_1 - operator of formation of tuple λ, calculation of initial values of $\tilde{L}_q$ and F_{cur}, assignment $i := 0$; A_2 - operator of main element selection, organization of the assignment cycle $i := i+1$; $j := i$; A_3 - operator of complementary element selection, assignment $j := j+1$; A_4 - operator of (i, j)-rearrangement testing, calculation of new $\tilde{L}_q$, $F(i,j)$, $\Delta F(i,j)$ values; A_5 - operator of calculation of $\Delta B(i,j)$; A_6 - operator of (i, j)-rearrangement and $\tilde{L}_q$ fixation, assignment $F_{cur} := F(i, j)$.

Logical conditions $w_i = 0$ correspond to following situations in pair arrangements:

w_1 — conditions 1 and 2 are fulfilled;

w_2 — $\Delta F(i, j) \geq 0$;

w_3 — $\Delta F(i, j) = 0$;

w_4 — $\Delta F(i, j) \leq 0$;

w_5 — $j \geq 0$;

w_2 — $i \geq N - 1$.

When $w_i = 1$ — jump by pointer. This algorithm has limiting ACC of about $O(0.5n^2)$. In some cases, because of limitation to element mobility during rearrangements, the optimal result cannot be reached by a single use of the algorithm, and the solution process should be repeated.

3.10 Striping of Combined Topology

Technologically invariant CAD system; conflict situation; interlayer connection; combined topology.

The currently existing CAD systems are oriented at definite LSI production techniques. The main disadvantages of such systems are:

- the impossibility of using the same CAD for LSI design with different or combined technology;
- the necessity of developing and using a new design system when LSI production technique changes and, as result, of personnel retraining.

In this connection it is very important to develop algorithms and systems can address the situations of change from one technique to another. Such design systems allow for performing whole system configuration for a required variant of LSI production and are meant to be parts of *technologically invariant* (TI) LSI CAD. TI CAD is a CAD system which allows for selecting the whole of machine programs proper for a design object or object class, considering their production technique.

In solving the combined topology striping problem for connecting LSI elements in TI CAD one must consider the *construction and technological limitations* (CTL), defined by LSI production technology, which change when LSI production technique changes. Such limitations can concern:

- the number of layers, where connection fragments can be allocated;
- the set of CTL, defining vertical LSI structure for a given production technique (the vertical structure of LSI is the structure of integrated circuit profile);
- the priority of layers, motivated by their conductivity, and, consequently, the requirement of allocating in layers with the highest conductivity of the longest connections fragments;
- the density of tracks of such layers ought to be maximal;
- the characteristics of LSI topology, defined by production technique;
- the metric parameters of track fragments that are dynamic and can change depending on allocation in definite layer.

Besides these requirements, in solving the combined topology LSI striping problem we must find, analyze and remove the situations of potential conflict, caused by the variety of possible arrangements of connection fragments on the plane and the CTL set of vertical structure of LSI.

The analysis of LSI production technology shows that in practice 4 to 13 layers (masks) are used, and it follows from the analysis of different LSI realization variants that this suffices for realization of the existing chip types. A general description of admissible and restricted combinations of layers in chips, realized using a set of masks, is provided by construction and technological basis (CTB). For the combined topology LSI striping problem, CTB can defined as relation $\psi_3^* = < C_3^*, B_3^* >$. The set of relation elements $B_3 = B^* \cup B_2^*$, B^* being the set of base layers, $|B| = n^*$; and B_2 - the set of possible interlayer connections $|B_2^*| = n_3^*$. If elements $b_i, b_j \in B^*$, then they satisfy graph relation C_3^*, i.e. if $<b_i, b_j> \in C_3^*$, then combination of layers of b_i and b_j is feasible. If $<b_i, <b_j, b_k>> \in C_3^*$, where $b_i \in B^*$ and $<b_j,b_k> \in B_2^*$, then combination of layer b_i and interlayer connection from b_j to b_k is feasible. If $<<b_i, b_j> <b_l, b_k>> \in C_3^*$, where $<b_i, b_j>, <b_l, b_k> \in B_2^*$, then in the vertical LSI structure it is allowed to use in the same coordinates the interlayer connections $<b_i, b_j>$ and $<b_l, b_k>$ simultaneously.

Assume that T, $|T| = k_1$, is the base set of technological variants of LSI production. For a given CTL set of LSI vertical structure and selected t_j^{th} technological variant of production ($t_j \in T$), it suffices to define relation $\psi_3 = < C_3, B_3 >$, $C_3 \subseteq C_3^*$, $B_3 \subseteq B_3^*$. Here $B_3 = B \cup B_2$, where B is the set of layers, used in

t_j^{th} technology, $B \subseteq B^*$, $|B| = n$; B_2 is the set of feasible interlayer connections: $B_2 \subseteq B_2^*$. B_1 is the set of layers used in t_j^{th} technology for commutation, $|B_1| = n_2$, $B_1 \subseteq B$; H ($|H| = m_1$) is the set of connection track fragments, and H' ($|H'| = d$, $H' \subseteq H$) is the set of connection fragments cross each other. Assume that a connection fragment $h_i \in H'$ is in layer b_p, $h_i \in b_p$, $b_p \in B_1$. Layer b_p is associated with the layer set D_p ($\Gamma(\{b_p\}) = D_P$), $D_p \subset B$, so that composition of b_p is possible with any layer b_k ($b_k \in D_p$) as allowed by LSI vertical structure, i.e. condition $(\forall b_k \in D_p)$ $(<b_p, b_k> \in C_3)$ is fulfilled. In this case the *combined-topology LSI striping problem* is formulated in the following way:

We look for such a division of the set H' into m subsets ($1 \le m \le n_2$), that the following conditions are fulfilled:

$$(\forall H_i \in H')\ (H_i \subset H' \wedge H_i = \varnothing);$$

$$(\forall H_i, H_j \in H')\ (H_i \ne H_j \wedge H_i \cap H_j = \varnothing);$$

$$\bigcap_{i=1}^{m} H_i = \varnothing;\ \bigcap_{i=1}^{m} H_i = E_1 \rightarrow E_1 = H';$$

$$(\forall h_i, h_j \in H_i)\ (h_i \ne h_j \rightarrow h_i \cap^* h_j = \varnothing);$$

$$(\forall H_i, H_j \in H)\ (H_i \in b_p \wedge H_j \in b_k \wedge b_p \ne b_k \wedge b_p, b_k \in B_1);$$

$$(\forall b_p \in B_1)\ (\exists \Gamma\ (\{b_p\}) = D_p \wedge D_p \subset B \wedge b_k \in D_p).$$

Here, $\cap^*$ means topological crossing of connection fragments h_i and h_j.

The combined LSI topology striping problem can be reduced to the problem of coloring graph vertices with a definite number of colors.

The model of topology of the combined LSI element connections, reflecting allocation of connection fragments on a plane, is constituted by an *orthogonal-diagonal graph* $G = (X,U)$. The set of vertices X ($|X| = C$) corresponds to the set of points, where connection tracks turn, and of final connection points. The entire set of plane tracks is examined as a set of connection fragments, into which these tracks are divided at turning points. The set of edges U is the set of connection track fragments, $U = H$, $|U| = |H| = m_1$.

An example of a fragment of combined-topology CMOS LSI is shown in Fig. 3.36. The corresponding graph is presented in Fig. 3.37.

To model intersections of edges of graph G on a plane, we use the intersections graph $G' = (X', U')$. The set of vertices in G' is equal to H'.

Any two vertices a and b in G' are connected with an edge $(a, b) \in U'$ if condition $(\forall\ (a, b) \in U')\ (a, b \in X' \wedge a, b \in U' \wedge a \cap^* b \ne \varnothing)$ is fulfilled, where $\cap^*$ denotes intersection of edges on a plane in graph G. Each vertex $x_i \in X'$, $i = \overline{1, d}$,

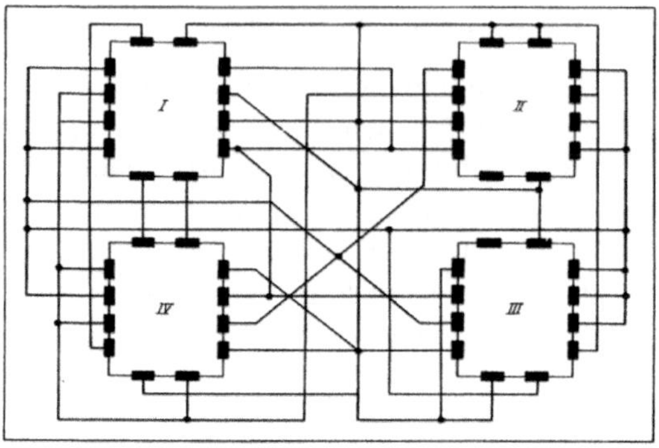

Fig. 3.36. Example of a fragment of combined-topology COMS LSI

has in *G'* the number of degrees of freedom equal the number of commutation layers in a given LSI production technology.

Assume the set of colors *A*, $A = B_1$, $|A| = |B_1| = n_2$. Each element $a_i \in A$, $i = \overline{1, n_2}$, is a color number (or color), corresponding to layer $b_i \in B_1$, $i = \overline{1, n_2}$, i.e. index of layer b_i. Coloring of vertices $x_e \in X'$, $e = \overline{1, d}$, of graph *G'* in one color $a_i \in A$, $i = \overline{1, n_2}$, leads to allocation of the same-label connection fragments $h_e \in H'$, $e = \overline{1, d}$, in corresponding layer $b_i \in B_1$, $i = \overline{1, n_2}$. Let $x_e^{a_i}$ denote vertex $x_e \in X$ colored with $a_i \in A$. Then, there is a set D_i of colors, with which vertex $x_k \in X'$, adjacent to vertex x_e of graph *G'*, can be colored.

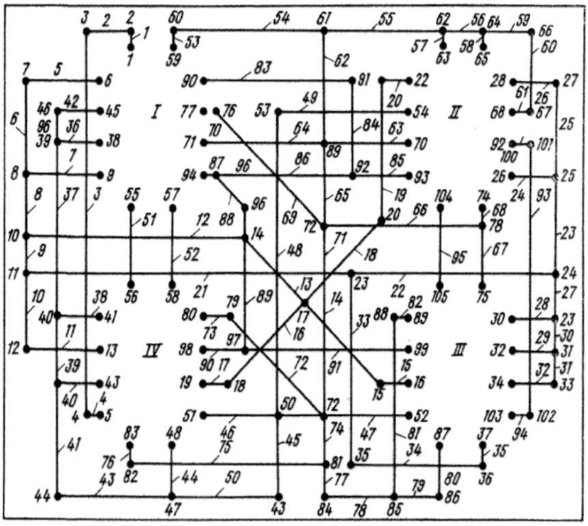

Fig. 3.37. Graph of the fragment of topology from Fig. 3.36

Striping of the combined-topology LSI into m layers is possible when there exists such coloring of vertices of graph G' with m colors that no two adjacent vertices have the same color and the following conditions hold:

$$(\forall a_i \in A)\,(\exists \Gamma(\{a_i\}) = D_i \wedge a_i \notin D_i \wedge D_i \subset A);$$

$$(\forall\, (x_e, x_k) \in U')\,(x_e, x_k \in X' \wedge x_i^{a_e}, x_j^{a_k} \in A \wedge x_i^{a_e} \neq x_j^{a_k} \wedge x_i^{a_e} \in D_j \wedge x_j^{a_k} \in D_i);$$

$$\bigcup_{i=1}^{m} \{x_e^{a_i}\} = X' \; ; \; 1 \leq m \leq n_2.$$

Structural diagram of a combined topology striping subsystem, meant for inclusion in a TI LSI CAD system, is shown in Fig. 3.38.

Information input of the combined topology LSI striping subsystem includes the following data: the base set (T) of variants of LSI production technology; construction and technological basis; rules of definition of the layers used in LSI: array of functional assignments of layers; the technologically oriented data.

The source information for solving the combined topology LSI striping problem is constituted by the number of technological variants (t_j) and the respective orthogonal-diagonal graph G.

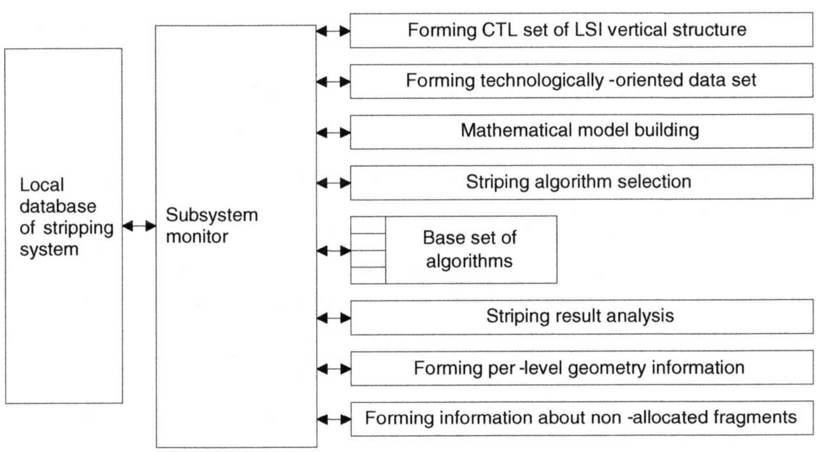

Fig. 3.38. Striping subsystem structural diagram

The striping subsystem can be divided into two main parts. In the first part the questions of system configuration for required technological variant t_j of LSI production are solved. Configuration is determined in two stages. The first stage consists in selection of information, required for solving the striping problem of combined topology of LSI produced with t_j^{th} technology. This selection of information proceeds as follows. The CTL sets of vertical LSI structure and the set of technologically-oriented data are determined. These data contain characteristics of topology. Based on analysis of graph G and selection of technologically-oriented data, graph G' of track fragment intersections is built. In the second configuration stage selection (definition) of algorithm f_k for striping track fragments in layers and its choice from the base set F^* are done. In selection of striping algorithm, two approaches are implemented, depending on the number of commutation layers in LSI, defined by t_j^{th} production technology. If IC production technology with one layer is selected, the approach, based on determination of planarity of electric connections graph with subsequent construction of its flat layout, is used. Direct selection from the base set F^* of algorithms is made by the subsystem monitor.

In the second part of the combined topology LSI striping subsystem, connection track fragments are divided into layers and level-related topological information is formed and, if required, information about connection fragments not allocated in layers. In division of track fragments into layers, coloring of graph G' with a given number of colors, is used, including the CTL set that defines LSI vertical structure for t_j^{th} technology of its production. In this context a comprehensive criterion is used for selection of a new vertex in G' and its coloring, through search from the best paths. The criterion is:

$$P_{b_j}^{h_i} = (kL + \beta N_f)hN . \qquad (3.16)$$

Here, $P_{b_j}^{h_i}$ is the weight of connection fragment $h_i \in H$ when it is allocated in layer $b_j \in B_1$; L is length of connection fragment h_i; N_f is the number of connection fragments removed from layer b_j when fragment h_i is allocated in it; N is priority of layer b_j; k, β, h are weight coefficients. The most suitable layer b_{j^*} ($b_{j^*} \in B_1$) for allocating h_i in it is the one with $P_{b_{j^*}}^{h_i}$ = min. This criterion allows for defining length of fragment h_i and the set of connection fragments $H' = H' \backslash \{h_i\}$ for each allocated connection fragment.

The process of dividing track fragments into layers consists of four parts.

Assume that connection fragments $h_j, h_k \in H', j \neq k$, form a turn at point v, i.e. v will be the point of incidence for them. A connection fragment that creates conflict situation (CS) in v, is h_i ($h_i \in H'$) crossing h_j and h_k at v, where interlayer connection can be allocated. The set of vertices F_5^*, where CS may occur between a connection fragment and interlayer connection, is defined as $F_5^* = \{x_j \in X / \alpha (x_j)\}, j = \overline{1, c}$. Hence, the vertex set F is defined by the expression $F = F_4^* \cup F_5^*$, $F \subset X$. The four parts mentioned are:

1. Determination of the set F of vertices (nodes) of graph G, where conflict situations can occur, arising from the set of fragments, allocated on the plane, and CTL set of LSI vertical structure, $|F| = m_2$.
2. Elimination of vertices $x_i \in F$, $i = \overline{1, m_2}$.

3. Determination of the set E of track fragments to be allocated in layers, $E \subseteq H'$, $|E| = p$.
4. Distribution of connection fragments $h_i \in E$, $i = \overline{1, p}$ among layers. As coordinates of interlayer connections are not known in advance, it is assumed that they lie in vertices of graph G. If some vertices in G have the same location on the plane, then the respective point has property α'. Then, the set of vertices F_4^*, where CS can occur between several interlayer connections, is defined as $F_4^* = \{x_i \in X / \alpha'(x_i)\}$, $i = \overline{1, c}$.

Assume that graph G has edges u_j, $u_j \in U$, $j \neq k$, incident to vertex v ($v \in X$) with coordinates (S_v, t_v), located on the plane. Assume that there is also in G an edge $u_i \in U$ ($i \neq j \neq k$), for which condition $u_i \cap^* u_j \neq \varnothing \wedge u_i \cap^* u_j \neq \varnothing$ holds. If any two edges of graph G cross on the plane, they have a common point. Assume that point w with coordinates (S_w, t_w) is the point where edges u_i and u_j cross (u_i, $u_j \in U$), and point n with coordinates (S_n, t_n) is the crossing point of edges u_i and u_k (u_i, $u_k \in U$).

So, for the examined graph G of Fig. 3.37 we obtain set $F = \{14, 15, 17, 20, 23, 72, 79, 97\}$. Concerning the set of vertices $F_6 = \{14, 20, 72\}$, $F_6 \subset F$, the edges, leading to CS, will cross the edges, incident to them, directly in these vertices. For example, the coordinates of crossing points of edge ~47 that cause CS in vertex 72, with edges ~71, ~72, ~74 will match the coordinates of point 72 on the plane (see Fig. 3.37), as described before. It is for the set of vertices at distance l_1, $l_1 < l$, where l is minimum admissible distance between the vertex, where interlayer connection can be placed, and connection fragment that causes CS. Thus, distance l_1 between vertex 15 and edge ~81 is smaller than l, ($l_1 < l$), and so it is considered that edge ~81 crosses edges ~14 and ~15 in a vertex, resulting in a CS in this vertex.

In the second part of the algorithm, considering CTL of LSI vertical structure, elimination of possible CS in vertices $x_i \in F$, $i = \overline{1, m_2}$ is performed. The way it is done is shown below.

1. Select next vertex $x_i \in F$, $i = \overline{1, m_2}$. Go to 2. After the whole set F has been examined, go to 6.
2. Distribute connection fragments causing CS in $x_i \in F$, among layers. If conflict situation is eliminated, go to 1, else, go to 3.
3. Eliminate conflict situations in $x_i \in F$ by moving interlayer connection to new coordinates. If such movement is possible, go to 1, else, go to 4.
4. Eliminate conflict situations in $x_i \in F$ by redistributing the previously arranged fragments in layers. If CS in x_i is eliminated, go to 1, else go to 5.
5. Fix connections fragments in CS in $x_i \in F$. Go to 1.
6. End.

Distribution of connection fragments causing conflict situations in a vertex x_i ($x_i \in F$) among layers, is done as follows. All the possible combinations, considering LSI vertical structure, of distribution among layers b_i ($b_i \in B_1$, $i = \overline{1, n_2}$) of connection fragments in CS, are generated, and weight of every combination is found. The weight of jth combination (P_j) is the sum of weights of connection fragments it consists of. Of all the available combinations, the most suitable is the one with $P_{j*} = \min_j \{P_j\}$. In general case, if there is a set S of connection fragments causing conflict situations, the problem of generating all possible layer combinations, where connection fragments $h_j \in S, j = \overline{1, n_3}$, can be placed, is equivalent to complete enumeration, and so is inefficient. To spare time and avoid reviewing all distribution variants, we suggest the following algorithm.

1. Form set S of connection fragments in conflict situation at vertex $x_i \in F$.
2. Select next layer b_j ($b_j \in B_1$) for locating the connection fragment $h_i \in H'$ causing conflict situation at x_i. Go to 3. If there is no layer b_j, go to 7.
3. Select connection fragment $h_l \in S$, $h_l \neq h_i$. Go to 4. If S has been entirely reviewed, go to 2.
4. Determine a suitable layer $b_{k*} \in B_1$ to locate h_l in it, $P_{b_{k*}}^{h_l} = \min_{b_k} \{P_{b_k}^{h_l}\}$.
5. Establish the possibility of combining layers b_j and b_{k*} in LSI vertical structure. If layer combination is possible, go to 6, else go to 4.
6. Define the weight of jth combination: $P_j = P_{b_j}^{h_i} + \sum_{i=1}^{n_3 - 1} P_{b_{k*}}^{h_l}$.
7. Select the best j^{th} combination for connection fragments distribution in layers: $P_{j*} = \min_j \{P_j\}$.
8. End.

The idea of this algorithm is that a connection fragment $h_i \in H'$ causing conflict situation in $x_i \in F$, is sequentially allocated in all layers admissible in view of its degrees of freedom, $b_j \in B_1$. For every placement of h_i in a layer b_j ($h_i \in b_j$), suitable layers $b_{k*} \in B_1$ are found for locating in them the connection fragments from the set S', where $S' = S \setminus \{h_i\}$, $|S'| = n_3 - 1$. So, the number of examined combinations will be equal to the number of degrees of freedom of fragment $h_i \in S$; this reduces the area of search for the most suitable solution and allows for reducing time of resolving the whole conflict situation problem.

Elimination of conflict situations in $x_i \in F$ by moving interlayer connections existing there to new coordinates, consists in the following. For edges u_j^i, $u_m^i \in U$ of graph G, incident to x_i ($x_i \in X$), coordinates of points, incident to vertex x_i, are determined. If the required interlayer connection can be located in these points, then it is moved to these points and CS at x_i is eliminated. For example, during CS elimination in some vertices of graph G of Fig. 3.37, it occurred that a connection fragment, corresponding to edge ~72 of G, was located in the diffusion layer. The connection fragment, entering the approximating rectangle IV and corresponding to edge ~73 of G, should definitely be located in the first layer of metal. In this case, a CS in vertex 79, as connection between the first metal layer and the diffusion layer is impossible there. The connection fragment corresponding to edge ~89 is located in the second metal layer. Then, to eliminate the conflict situation in vertex 79 of G, the coordinates of the point, incident to vertex 79 are defined on edge ~73, and interlayer connection of the first metal layer and the diffusion layer is moved from vertex 79 to the point with found coordinates, and so CS in vertex 79 of graph G is eliminated (see Fig. 3.37).

The third stage of the combined-topology LSI stripping algorithm consists in determination of the set of connection fragments to be distributed among layers. The set E of the connection fragments not allocated to layers corresponds to the set F_1 of the non-colored vertices of graph G', $E = F_1$, $|F_1| = P$, $E \subseteq H'$, $F_1 \subseteq X'$. Assume that E_1 is the set of connection fragments distributed among layers, and F_2 is the set of colored vertices of graph G', $E_1 = F_2$, $|E_1| = |F_2| = P_1$. Assume that a connection fragment $h_i \in H'$ has property α_1, if it is allocated in a certain layer $b_i \in B_1$ and vertex $x_i \in X$, corresponding to fragment h_i also has property α_1, if it is colored in color of layer b_i. In this case, $F_1 = \{h_i \in H' \,/\, \alpha_1 (h_i)\}$ and $F_2 = \{x_i \in X' \,/\, \alpha_1 (x_i)\}$, $i = \overline{1, d}$. On the basis of the above, the set of connection fragments not allocated among layers is determined, $E = H' \backslash F_1$, as well as the set of not colored vertices $F_1 = X' \backslash F_2$.

In the fourth stage of the combined-topology LSI striping algorithm division of connection fragments $h_j \in E$, $j = \overline{1, p}$, between layers, takes place.

Each vertex $x_j \in X'$, $j = \overline{1, d}$, corresponds to its set of degrees of freedom, L_j ($L_j \subseteq B_1$, $|L_j| = m^*_j$). This set defines in which colors can vertex x_j be colored, i.e. in which layers $b_i \in B_1$, $i = \overline{1, n_2}$, the connection fragment $h_j \in H'$ can be placed.

Depending on definite requirements, any of the examined graph coloring algorithms with time complexity from $O(n)$ to $O(n!)$ can be selected here.

The result of the combined-topology LSI striping algorithm is description of level-related LSI topology, containing following information: number i of a connection fragment $h_i \in H$, $i = \overline{1, m_1}$; coordinates of vertices, incident to edge $u_i \in U$, representing the connection fragment h_i; coordinates of interlayer connection allocation; metric parameters of topological description (width of connection fragments; admissible distances between connection fragments, and between connection fragment and interlayer connections; sizes of interlayer connections). If there are connection fragments that could not be allocated by the algorithm in the predefined layers, information is provided containing number j of the non-allocated fragment h_j ($h_j \in H'$), and coordinates of vertices, incident to edge u_j ($u_j \in U$), representing fragment h_j in graph G.

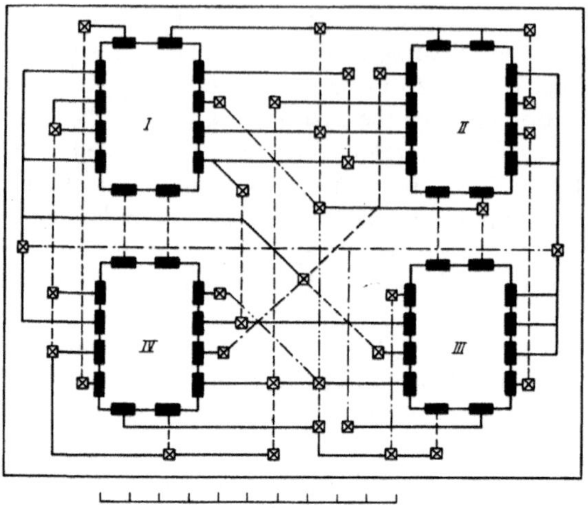

Fig. 3.39. Striping of a fragment of topology (see Fig. 3.36)

The effect of the combined-topology LSI striping algorithm, represented as level-related topological outline for the example of Fig. 3.36, is shown in Fig. 3.39. The complexity of suggested combined-topology LSI striping algorithm is $O(m_1^2)$, where m_1 is the number of connection fragments.

3.11 Description Language and Control of LSI Scheme Topology

Topological outline; level-related geometry; technical limitations control; modeling graph; topology description language.

The process of LSI photomap production is complex and requires high precision. Consequently, it is very important to develop the design system for LSI photomap design and production (LSIP CAD). Fig. 3.40 illustrates the structure of LSIP CAD system including the following stages: coding of LSI topological outline; information control; level-related LSI geometry outline development; control and analysis of the obtained variant of topology; obtaining control information for LSI photomap production equipment.

Output document from this LSIP CAD is topological outline of the LSI elementary diagram. As determination of topological outline of LSI is a laborious process, from which large numerical arrays are obtained, it is impossible to prevent all errors during first encoding. Errors, occurring in topological outline description, can be of two kinds: design and instrumental, caused by equipment errors. When description of topological outline originates from semiautomatic devices, errors of both kinds are possible. To find these errors, LSIP CAD systems use a syntactic control unit. Syntactic control program is the first unit of topological description language translator for photomap production equipment. The second unit of this

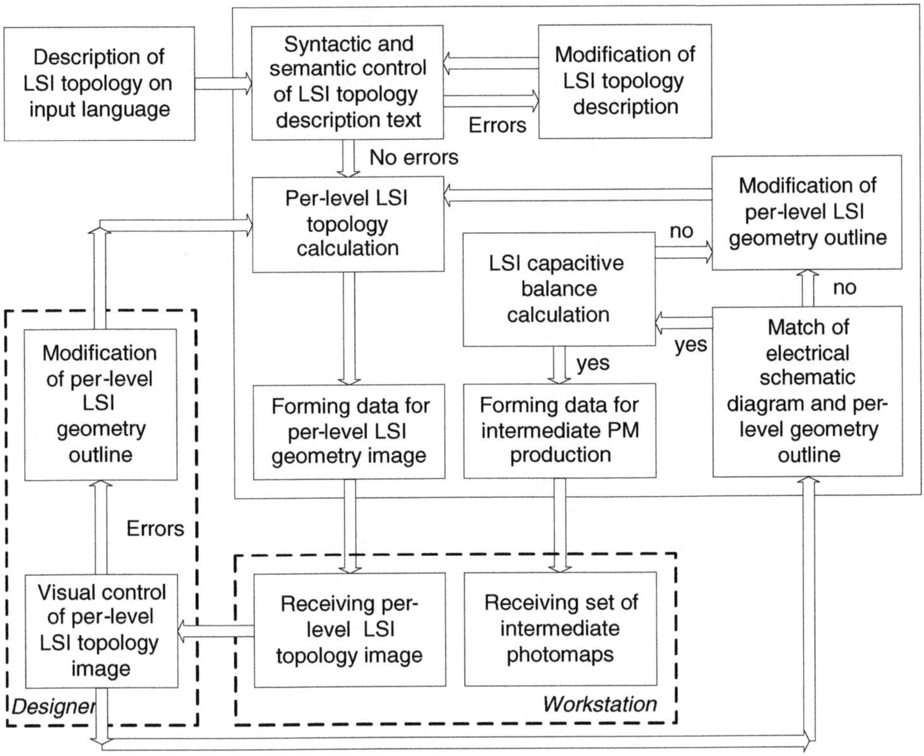

Fig. 3.40. Structural diagram of automated design of LSI photomaps

translator produces information on level-wise geometry, i.e. configuration of all elements on each level, considering element center coordinates and their structural dimensions. The results of level-wise geometrical outline calculation are completed with codes of respective device that traces the LSI level-wise geometrical outline. The outline obtained is analyzed by the designer. If the results are not satisfactory, the outline is corrected. Otherwise, automatic LSI topology control is carried out.

Control and analysis of topology includes control of technological limitations that ought to be controlled, of the electric source scheme of LSI and of the list of required electric parameters. This data is fed to the program and its syntactic control is performed.

Controlling of technological limitations consists in checking distances between areas in topology, their sizes and overlaps, and is done by the computer. Output information on violation of limitations is provided as text and/or outlines of areas that do not meet the limitations. The next stage consists in reconstruction of schematic diagram using available variant of topology. In this stage, incorrect connections of components are excluded by comparing the obtained circuit list with schematic diagram. If errors are found, LSI topology is modified. Then, using the schematic diagram obtained, equations, describing LSI working logic, are automatically formed. Control of the working logic of the scheme enables ensuring that all LSI connections are correct and the scheme fulfills its purpose.

Fully automatic control is important not only for the initial and final design data, but also for the intermediate ones. It allows for finding design errors on early stages, reduces time and cost of LSI design. Availability of fully automatic control of obtaining topological variants increases LSIP design system automation level and allows for production of correct photomaps in one step. Usually, in LSI design it is attempted to keep track of element labels, as this simplifies the control process. Keeping element labels in all stages gets difficult when designer interacts with computer using display and languages for LSI topology description. Then, the problem of identification without elements labeling arises. Computer use enables speeding up the scheme control process, decreases its cost and increases reliability of results. Processing of schemes without element labeling can be done using the algorithm of finding isomorphism between LSI and graph substitution. For this purpose, of course, scheme is modeled with a graph.

The respective *model graph* is non-oriented with separation of the set of vertices. Introduction of separation broadens graph modeling capabilities. Two subsets of LSI elements are selected. On these subsets relations are defined. Then, next subset of elements is introduced and new relations between this and previous subsets are defined. This process continues until an exact scheme description is obtained. Then, the elements are replaced with vertices, and relations with edges of bipartite subgraphs. A graph is constructed. Different kinds of elements define the separation of graph vertices. Graph with separation is equivalent to scheme description. By selecting different scheme descriptions, we can obtain different model graphs. Representation of relations by bipartite subgraphs and use of separation of vertices enable accounting for the features of the scheme. The schemes are represented by model graphs and checked for isomorphism. If graphs are isomorphic, then isomorphism substitution defines the correspondence between the elements of the scheme. If graphs are not isomorphic, two schemes considered are not equivalent. The model graphs of LSI are constructed using electric and topological schemes.

Let us examine the functioning of the algorithm on an example. Assume two graphs: $G = (X, U)$, a model of LSI CD, and $G' = (X', U')$, a model of topological scheme, $|X| = |X'|$ and $|U| = |U'|$. Assume an extreme case, when all local degrees of graphs G and G' are mutually equivalent. We look for a substitution, if it exists, for which $t(X) = X'$, $t(U) = U'$, and then graphs G and G' will be isomorphic.

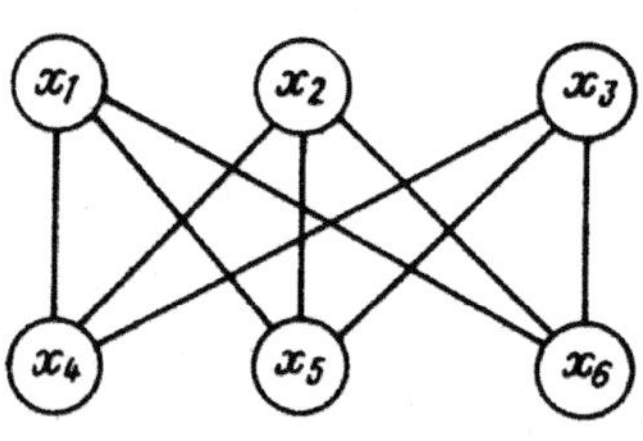

Fig. 3.41. Graph G (full bipartite graph)

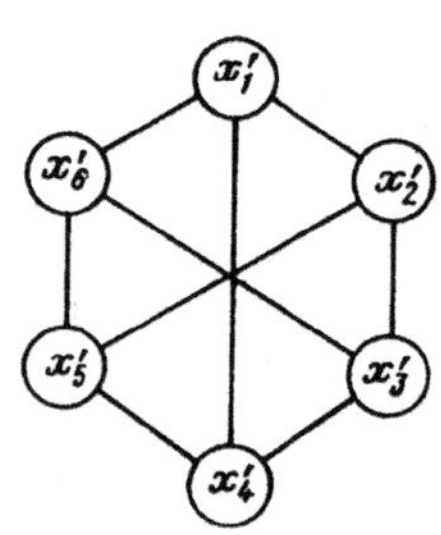

Fig. 3.42. Graph G', isomorphic to graph G (Fig. 3.41)

Take graph G of Fig. 3.41 with $|X| = 6$, $|U| = 9$, and graph G', $|X| = 6$, $|U| = 9$. Assume that vertices x_1 and x_1' are isomorphic, and in relation to them we divide all other graph vertices to adjacent and not adjacent subsets:

$(x_1)\ (x_2, x_3)_0\ (x_4, x_5, x_6)_1,$

$(x_1')\ (x_3', x_5')_0\ (x_2', x_4', x_6')_1.$

Then, the presence of edges between vertices x_2, x_3 and x_3', x_5' is checked. It can be seen in Figs. 3.41 and 3.42 that there are no such edges. The subset of smaller cardinality is separated, i.e. (x_2, x_3). Select vertices x_2 and x_3 and assume that they are isomorphic and then separations lines can be written as follows:

$(x_1)\ ([x_2]^0, [x_3])_0\ (x_4, x_5, x_6)_1,$

$(x_1')\ ([x_3']^0, [x_5'])_0\ (x_2', x_4', x_6')_1.$

It can be deduced there from that vertices x_3 and x_5' are, as expected, isomorphic (E-isomorphic). Separate subsets (x_4, x_5, x_6) and (x_2', x_4', x_6'). Assume that vertices x_4 and x_2 are E-isomorphic. After separation we get:

$(x_1)\ ([x_2]^0, [x_3])_0\ ([x_4], ([x_5]^0, x_6)_0)_1,$

$(x_1')\ ([x_3']^0, [x_5'])_0, ([x_2'], ([x_4']^0, [x_6'])_0)_1,$

which defines isomorphic substitution:

$$t = \begin{pmatrix} x_1 & x_2 & x_3 & x_4 & x_5 & x_6 \\ x_1' & x_3' & x_5' & x_2' & x_4' & x_6' \end{pmatrix},$$

that converts graph G to graph G'. Time complexity of the algorithm ranges between $O(n^3)$ and $O(n!!)$, with the most complex solutions corresponding to regular graphs that are not isomorphic.

Note that while developing software systems for schematic diagram reproduction by LSI topology the following problems must be solved:

- selection of models for graphical information presentation in computer memory and development, on the basis of these models, of fast algorithms for processing of topological data
- selection of a model that describes search rules and reproduction using topology of elements and connections, that can be adapted to changes in technology.

During production of LSI photomaps the coordinatographs or micro-photographic typesetters are used, their respective control information containing about 10^{i+1} numbers, where i is LSI integration level. That is why this information is prepared using computers. The source data for the computer is constituted by the detailed description of LSI topology, which can be established using two methods: by setting all corner points of topology areas or using specially developed languages. Description of topology by coordinates of all corner points is significantly redundant. After exclusion of redundant data LSI topology can be defined correctly. Reduction of numerical information decreases encoding complexity and the possibility of errors.

Selection of the method for describing input information, input language and translator, determines in many aspects flexibility of the design system, speed of adaptation to construction information on a new variant, and serviceability. Main requirements on the *topology description language* are:

- problem orientation, allowing simplicity of use by designer of topology;
- securing description compactness for high-complexity schemes;
- applicability to a wide class of design and technological problems;
- independence of computing and hardware-controlled equipment type, used for outline and photomap production.

Considering these requirements, preference should be given to special topology description languages instead of description modeling using general purpose languages like Fortran.

Currently, there are a number of languages that meet these requirements to a certain extent. They have different syntax and capabilities. For example, the technology description language is oriented at a certain class of chips and symbolic design methods. Similar problems are solved by the Symbol language. Here below we provide the description of input language, used in the LSI topology CAD system.

The alphabet of this language consists of following symbols:

1. uppercase letters of the Latin alphabet;
2. ten digits, from 0 to 9;
3. seven special characters: * — asterisk; (— left parenthesis;) — right parenthesis; , — comma; ⌉ — end of sentence; " + " — plus; "–" — minus.

For the sake of simplicity, clarity and usability, syntax structure of input languages is used as a basis. Information is represented in arrays of the same-type objects. Each array has a name. List of information on array objects consists of structures like *A**K*, where *A* is array name; *K* — array object; * — separator. Structures in the text are divided with commas. Text can be divided into sentences of any length. End of sentence is marked by the separator ⌉, for example, *A**K*, *B**C*, *N**P*⌉.

Structures, following in sentences one after another and having same array name, can be written with obligatory specification of name for first structure. For the subsequent structures it is required to skip the name: *A**K*, *C*, *E**L*. Any object can be connected with objects of other arrays. All of them are written down directly next to the object they are linked to, in brackets:

*A**K*1 (*M**T*, *C*, *E**L*)⌉

*B**P*1 (*F**N*2 (*G**D*5))⌉.

In view of complexity of formal representations, semantic rules are often defined informally, in natural language.

Input chip topology information consists of the library of topological elements, library of scheme components topolog (typical cells), and functional scheme.

When developing the library of elements of topology, sizes of transistors, diodes, resistors and other elements of the scheme are defined, accounting for topological constraints. Here and later on for the names of arrays the first letters of respective words will be used. For example, *L* — layer, etc.

The following arrays are used for describing the library elements:

T — title (type, name) of the component;

L — number of topological layer;

H and *B* — respectively, height and width of shape;

X and *Y* — coordinates of left bottom vertex of shape in this layer.

As an example, we will describe allocation and sizes for the topological layer 1 of an *n-p-n* transistor with Shottky diode *VT*1, produced using technology with isolation by *p-n* junctions. *VT*1 topology is shown in Fig. 3.43, where 1—7 are layer numbers.

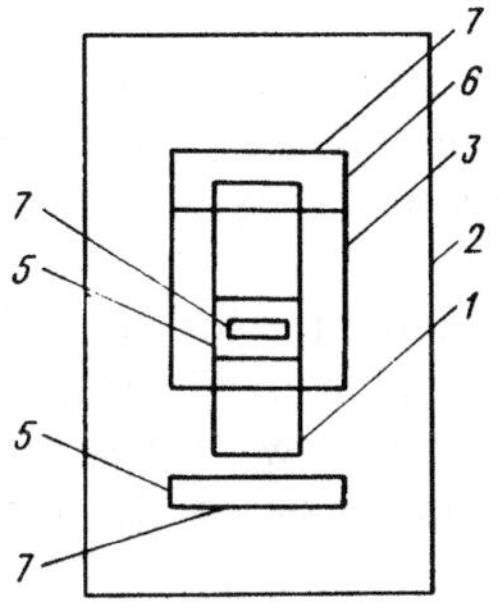

Fig. 3.43. Example of topology of transistor VT1

Coordinates are calculated from the left bottom corner of transistor:

*T**VT*1 (*L**1 (*H**33, *B**12, *S**15, *t**15))⌉

In the same way the entire transistor is described:

*T**VT*1 (*L**1 (*H**33, *B**12, *S**15, *t**15);

*L**2 (*H**63, *B**42, *S**0, *t**0);

*L**3 (*H**21, *B**20, *S**11, *t**23);

*L**5 (*H**4, *B**20, *S**11, *t**10, *H**7, *S**12, *t**15, *Y**27);

*L**6 (*H**7, *B**20, *S**11, *t**44);

*L**7 (*H**4, *B**20, *S**11, *t**10, *H**1, *B**6, *S**18, *t**30, *H**7, *B**20, *S**11, *t**44))⌉

There is a possibility of shrinking (extending) a typical element along the axes *S* and *t*. For example, entry *T***VT*1 (*IX**20) means that the library element *VT*1 should be "extended" along *S* axis by 20 units.

If one of the topological layer elements is a "ring", then first the coordinates and sizes of external contour are defined, and then, using symbol " + " – of the inner contour, for example

*L**2 (*H**40, *B**20, *X**15, *Y**15, +, *H**10, *B**10, *X**20, *Y**30)⌉.

There is a possibility of building arcs or circles. To do this, we have to define coordinates of circle center and its radius, coordinates of initial point (arrays *S*1 and *t*1) and of the end of an arc (*S*2, *t*2). An example for arc with center in point (0; 0) and radius 1:

*L**2 (*S**2, *t**0, *S*1*1, *t*1*0, *S*2*0, *t*2*1)⌉.

Typical library of fragments consists of element allocation and connections between them. Here, positions of elements and layer metallization topology are defined. There is a possibility of using library elements in different orientation and their multiplication by *s* and *t* axes. An extra array *p* corresponds to an enlarged cell. It is created when in one enlarged file a part of scheme topology of typical cells is provided, but multiplication cannot be used.

The disadvantages of this language are the necessity of having the topological outline and data volume increase for VLSI topology design. The main advantages of the language are simplicity, possibility of use for describing a wide class of digital LSI topologies, and complete independence of the language of element base and production technology.

3.12 Questions of Hardware Support of VLSI CAD Software

Hardware support; knowledge base machine; engineering design machine; electronic simulation; specialized computational simulation device.

Most of the ECE LSI design problems are related to *NP*-complete or *NP*-hard problems. This makes the process of obtaining high-quality solutions, i.e. globally optimal or near to global optimum, significantly more difficult. The situation is even more complex, because these problems refer to a high number of criteria, a variety of requirements and limitations, and it is necessary to obtain their comprehensive solution.

The higher the level of the object designed, the lower the problem dimension, and vice versa. So, complexity of problem solving increases, as usual, exponentially, while the level of the object designed decreases. For design of high-level objects algorithms with exponential complexity can be used, but for objects of lower level these algorithms will take hundreds or thousands of years even on fastest universal computers to solve.

So, in problem of CAD improvement, time factor (systems performance) is very important. That is why along with structural improvement of the traditional CAD systems (using universal computers) simultaneously new approaches to CAD construction are developed. One of such approaches is improvement and extension of CAD hardware component, founded on selection of basic macro-operations, electronic simulation of problems and parallelization of data processing.

However, it should be remembered that it is important not only to gain high technical effect, e.g. performance, but also an appropriate cost-benefit relation. Hence, efficiency E of a complex system is expressed as the ratio

$$E = D/C, \tag{3.17}$$

where D is benefit from the new or modernized system; C is cost of system development, introduction and exploitation. A correctly selected efficiency index can help even in early design stages to assess a definite improvement of a complex system.

Efficiency is assessed by particular quality indexes (PQI) $y_i \in Y$, $i = \overline{1,n}$, and the generalized indices (GQI), being linear combinations of PQI:

$$E = F(y_1, y_2, \ldots, y_n) = \sum_{i=1}^{n} b_i y_i / C, \tag{3.18}$$

where b_i are weight coefficients.

When system efficiency is assessed with one PQI, the other ones are assumed to stay within defined bounds.

In CAD systems the most important PQI are: y_1 — performance; y_2 — reliability; y_3 — hardware complexity, and y_4 — system cost.

It is convenient to operate not with absolute, but with relative efficiency. So, in formula (3.18), instead of y_i and C, respectively, $\overline{y}_i = y_{id}/y_i$ and $\overline{C} = C / C_{\max}$, are used, where y_{id} is the ith PQI value for the system developed; $C_{\max}$ is the maximum admissible cost of the system. Now, GQI is equal

$$E = \sum_{i=1}^{n} b_i \overline{y}_i / \overline{C} \; . \tag{3.19}$$

In comparing various system variants it is expedient to have a fixed level of generalized (or particular) quality index, with which GQI (or PQI) values of variants are compared. This fixed level is preferably taken as equal 1. Hence, formula (3.19) should be divided by $\sum_{i=1}^{n} b_i$.

Then, GQI is equal 1 for $y_i = y_{id}$, $C = C_{\max}$, $i = \overline{1,n}$, and the formula for GQI is

$$E = \sum_{i=1}^{n} b_i \overline{y}_i / \overline{C} \sum_{i=1}^{n} b_i \tag{3.20}$$

with $\overline{y}_i = y_{id}/y_i \geq 1$, $\overline{C} = C / C_{\max} \leq 1$, $i = \overline{1,n}$.

The ratio (3.20) is the degree of purposefulness of system realization or modernization. So, if GQI, calculated with (3.20), is greater than one, the newly developed or modernized system will be efficient. Weight coefficients b_i can be determined using expert estimates.

Analysis of modern conditions and tendencies in VLSI CAD development shows that intensive work is done on the use of universal mainframe supercomputers like CRAY as technical CAD equipment. Moreover, special problem-oriented computers are intensively developed. Endowed with capacity of realizing any algorithm, universal computers cannot compete with specialized ones in solving, for instance, the engineering design problems. These problems are solved 2-10 times faster using hardware and software means than when only software resources are used.

Creation of fifth generation computers assumes the presence of three parts: hardware, software and external interface. Hardware resources should include machines for knowledge bases, dialoging, and problems solution.

Problems solution machines will provide hardware support for main algorithms in definite knowledge domains. The key question in VLSI CAD is hardware support for composition, allocation and routing algorithms.

We can suggest a system for electronic simulation of automated design problems (LSI engineering design machine, Fig. 3.45), oriented at solving not just one problem, but a whole complex of them. The system is based on units and devices (with hardware support of algorithms for main design problems). The system includes a model unit (MU) containing graph (or hyper-graph) topology representation and uniform routing environment; a uniform commutation environment (UCE); a control unit (CU) consisting of a clock generator, units for setting of initial conditions and limitations; an enumeration organization unit (EOU), and a solution organization unit (SOU). Using such system as a hardware support for the CAD system will allow for combining the advantages of the CAD system itself and the performance qualities of specialized processing units. It is assumed that functions are parallelized between hardware and software. The traditional CAD systems use hardware devices only for information representation, and all algorithmic processes are realized with software, but the new approach increases the role of hardware. Algorithms are realized now by specialized hardware devices enabling increase of performance several times over. Software provides information flow control, output of design documentation using special information representation devices, and effective changes of algorithms.

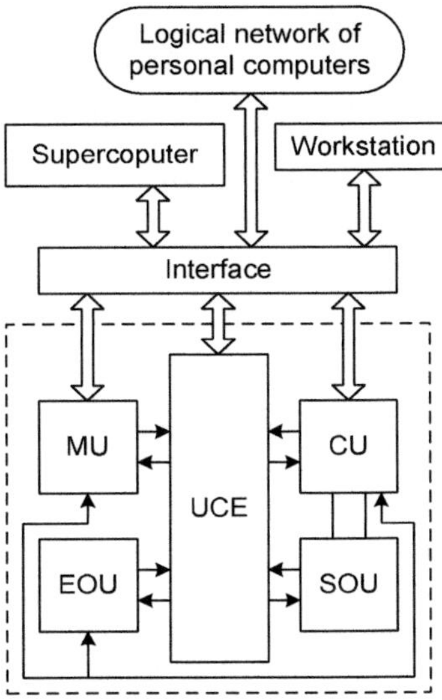

Fig. 3.44. Possible structure of CAD system hardware component

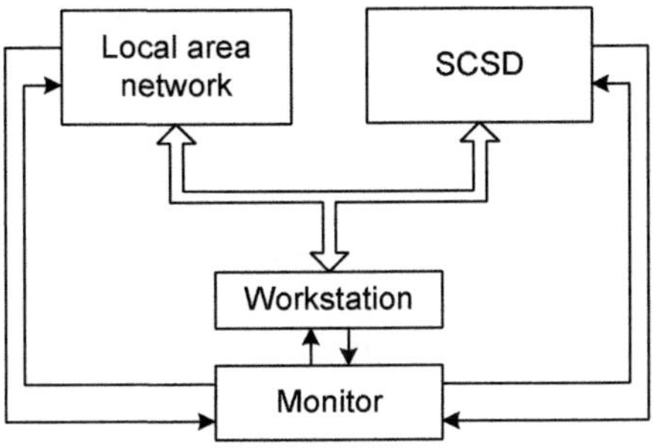

Fig. 3.45. Machine for VLSI engineering design (left diagram)

A possible structure of the hardware part of CAD is shown in Fig. 3.44. It contains a traditional workstation consisting of minicomputer, multiplexer, graphic information encoder, alphanumeric and graphical displays, plotter and coordinatograph. The hardware component includes also a specialized computation and simulation device (SCSD), local area computer network and monitor. SCSD secures hardware realization of the basic and auxiliary algorithms of the design process, listed above. The monitor ensures interaction between all parts of the system.

The main purpose of SCSD is realization of high-performance algorithms. Two problems are most important in this context. The first one is to find and develop algorithms most adapted for hardware realization (problem of adequacy of algorithms to hardware), the second is physical realization capacity. If we use the notion of hardware complexity of realization, as analogous to algorithmic time complexity, then "good" hardware realization should be the one, whose hardware costs grows, depending on problem dimension, in degree not more than 2.

Obviously, hardware realization of algorithms can give significant advantage in terms of performance only due to specialization. Thus, hardware realization of sequential composition algorithm can increase solution speed 10-100 times. Use of algorithms adequate to hardware and deep parallelizing can produce even more significant gains, especially for problems of large dimensions.

That is why random search algorithms are of interest. Attempts of using these algorithms started long ago, but they were based on slow generation of pseudorandom numbers by universal computers. Let us add that we do not need just plain random (or, rather, pseudorandom) numbers, but relations of random sets (random rearrangements, combinations, allocations) with constant numbers of elements (elements, and their changing orders). During generation, we should first obtain a sequence of pseudorandom numbers and then convert it into a random relation; this requires additional time and reduces performance.

In hardware realization there is a possibility of parallel generation of random connections. So, this appears to be the future of solving problems of high dimensionality. It is based on review of the set of random solutions Q, which, with at least a given probability P, contains solution R, differing from the optimal R_{min} by not more than some predefined ε, i.e. $P\{(R-R_{min}) < \varepsilon\} \geq P$.

In solving the allocation problem MU establishes CD MM and defines distances between positions in commutation field. EOU performs element selection and allocates elements in respective positions. SOU estimates overall connection length and memorizes the best allocation.

In solving the routing problem the hardware models are based on wave algorithms and their modifications. The hardware routing models use MU, UCE, EOU, SOU and the procedure of signal propagation from one cell to another. In this context the operation of construction of the shortest paths between definite cells is realized; cells simulate CD contacts. Cells are nowadays usually the data processors, microprocessors, personal computers, cells of uniform and quasi-uniform media. Because of almost instant signal propagation, the speed of algorithms increases abruptly. The main disadvantage of hardware models is sharp increase of the number of elements, and appearance of reliability and heat-radiation problems.

Note that realization of SCSD will require development and creation of new hardware, like composers of deterministic and random combinations, rearrangements, code compacting devices for parallel information processing, units for presentation of graphs, hyper-graphs, etc. SCSD can be used for routing, composition and other machine-building tasks, and also for hardware support of the VLSI CAD knowledge base. All this gives hope that in a near future the SCSD will appear that will make it possible to achieve a new quality level of LSI and VLSI design.

3.13 Design of LSI Topology Using Silicon Compilers

Silicon compiler; hierarchical editor; chip plan; cells merging; silicon assembler.

The necessity of designing specialized custom VLSI and the increase of their degree of integrity stimulate development of new automated design methods. The procedural design methods or *silicon compilers* (SC) are among them.

SC is a CAD system that realizes process of abstract computer or ECE topology description. It can be divided into three main parts:

- physical synthesis tools for creating crystal base plan, division of the scheme, allocation, routing, etc.;
- tools for logical synthesis of combination or sequential schemes;
- procedural design and unit generation tools (realized using so called silicon assembler programs).

An outline for the process of VLSI topology design is shown in Fig. 3.46.

The SC-based CAD systems consist of the following subsystems: source information input, crystal field planning, introduction of library elements, silicon compiler kernel, hierarchical editor.

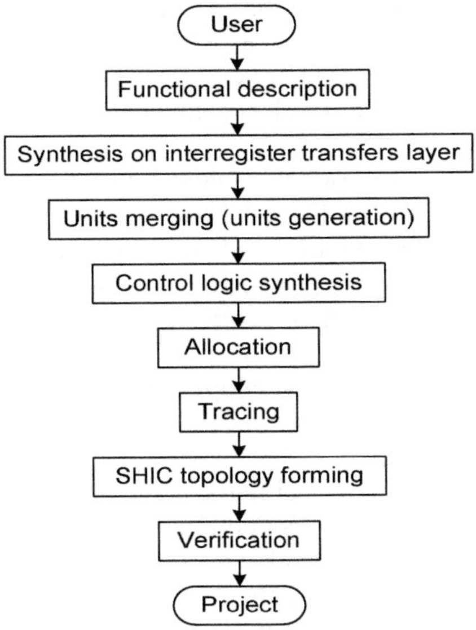

Fig. 3.46. Design of VLSI topology using SC method

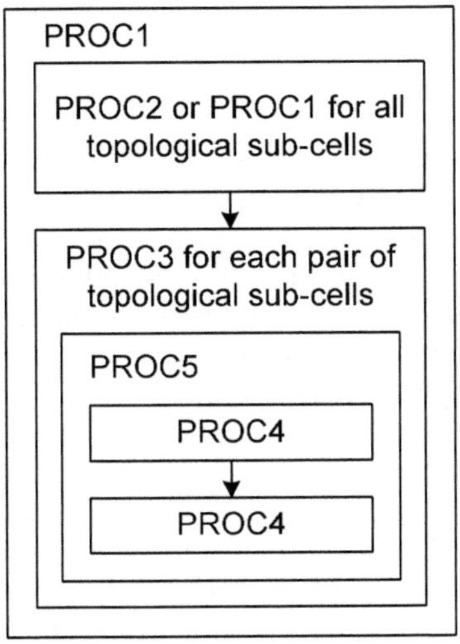

Fig. 3.47. Interconnection of procedures during cell merger

The silicon compilation process is realized by translation of input description into crystal topology. It consists of two cycles. In the first one, the level of source description of VLSI is lowered so that it can be interpreted by SC; in the second - the process of silicon compilation is realized. The first cycle is organized by the source description input subsystem. Input information is constituted by the functional diagram of the object designed in a high-level language. After input and translation of the designed object (DO), source description is analyzed. This analysis consists of two operations: top-down detailed elaboration of functional description and replacement of functional representation of cells with the logical one.

In control of the design process the *chip planning* system is used. A chip plan is a preliminary variant of VLSI cell allocation. It is developed using information on estimates of cell sizes. The output from the subsystem enters the central DB. After the two systems, examined above, produced their output, the intermediate design task is obtained as text description of the source data in EDIF language.

The SC kernel consists of three parts: central, cell merging algorithm, and silicon assembler. All the data about the intermediate DO state are stored in the central DB. The cell merging algorithm is considered to be the most important software procedure of SC. Its result is a cell design. The algorithm works recursively until VLSI crystal is designed.

The *silicon assembler* is a program that realizes cells of definite type. It is considered that silicon assemblers connect SC with a given set of technical requirements and limitations. Silicon assemblers consist of counters, registers, uniform medium cells, etc.

The standard elements library contains all descriptions of cells used for design, and information for input description analysis. There are tools for editing, description of library elements in graphical language, and adding new data.

When the design of a crystal does not satisfy the developer, hierarchical editor is used that allows for rebuilding of cell hierarchy.

VLSI topology design in SC consists of software-implemented connection between logic and topology. That is why, unlike in traditional CAD systems, functional description is brought to SC input at behavioral level. Input information shows what functions should be realized by the device designed, but not how these functions will be realized. This can be data flow graph, or textual record in special languages like ISPS, ICL.

Synthesis on inter-register transfer level consists in development of the structure of device, according to the obtained description. The units, required for realization of the scheme, should be defined at the output.

Coordination of units consists in definition of physical units, corresponding to units of preceding stage. Descriptions of physical units can be found in the library or are created by software (procedures) using macro-cell generation. Solutions developed in this step play the most important role and define constraints on system parameters (cost, area, etc.). After physical units are selected and coordinated, synthesis of control logic is made.

Like the traditional CAD systems, SC solves problems of topological synthesis using allocation and routing. Modern systems of topology synthesis in SC include stage-wise optimization of chip area and characteristics. Large dimensions of problems solved affect allocation and routing algorithms of SC. Sequential allocation and wave routing algorithms are considered as most suitable. The stage of topological synthesis remains one of main reserves for improving VLSI design quality.

Let us examine synthesis of topology on the basis of typical cells, called cell merging method. We refer to realization of this method for typical quadrilateral cells. A typical cell is a set of shapes (rectangles, polygons, conductors), belonging to certain layer, and four lists of ports (external outlets at each side) of a cell. A description contains the following information:

- coordinate values along axes s and t in rectangular coordinates;
- layer number;
- number of signal (circuit);
- width of conductor (connection);

Input information for synthesis of topology is hierarchical multilayer input description of VLSI enabling formation of lower-level typical cells and defining interconnections between typical cells of lower and upper levels. A typical cell of an upper level is a cell obtained after merging of some cells of lower hierarchical layer. These cells are described as topological sub-cells of the given higher-level typical cell.

The quality criterion for the topology of synthesized LSI is as follows:

$$F = \sum_{i=1}^{l} \lambda_i Q_i ,$$

where Q_i is reduced (normalized) estimate of ith quality index; λ_l is weight coefficient of ith index; l is the number of indices.

Let us now present the algorithm of automatic synthesis of VLSI topology using sequential allocation of typical cells.

Hierarchical input description allows for representing the whole process of VLSI topology synthesis as a simple, repetitive procedure of merging typical cells. The algorithm of synthesis of topology is as follows:

1. Define all topological cells.
2. If there is one topological cell, then go to 4, else go to 3.
3. Merge pairs of all topological sub-cells.
4. End.

In defining topological cells, one of two cases can occur: it is a typical lower-level cell; it is a typical upper-level cell. In the first case the topological sub-cell is defined by calling the procedure of lower-level typical cell formation according to a definite input description. In the second case it is defined by the recursive call of the described procedure.

Merging of typical cell pairs involves defining their positional relationship with the layout of all connections between them, including topology of all required transit tracks and power line layout in this area. As the result of merging the pairs of all topological sub-cells, a single typical cell is formed; it is a line of cells with topology of interconnections, transit tracks and power lines in areas between cells.

To prevent allocation of all typical cells of hierarchical input description in one line during topological sub-cell merging, each of them is rotated by 90°. Due to this, each topological sub-cell in a line is a column. Each typical cell, obtained from cell merger, can be treated as a topological sub-cell again and be merged with other cells.

The algorithm of merging two typical cells consists of:

1. Determination of the transit signal list for cells merged.
2. Determination of lists of signals, required to the left and to the right of the cells, and addition of transitive signals to them.
3. Formation of a typical cell containing topology of all tracks between cells.
4. Merging of typical lower cell with cell interconnections and layout of power lines between cells.
5. Merging of the resulting cell with typical upper cell and layout of power lines between them.
6. End.

The synthesis of VLSI topology proceeds from left to right on the plane.

From the area between cells merged tracks are laid, required for connecting with the already defined typical cells and typical cells to be defined later, and also tracks being interfaces. So, two lists of signals are formed that should appear between cells on the left or right side of the resulting cell.

Formation of interconnection topology can be done by any channel routing algorithm that allows for obtaining the layout of 100% of connections for unlimited channel width.

Merging of a lower typical cell with interconnection cell, and of resulting cell with the upper one is done so as to enable allocating cells at a distance from each other, this distance being defined by the necessary channel width, i.e. width of the interconnections cell.

Power line layout is performed separately from interconnection synthesis, as these lines are laid only in one layer (in metal). To guarantee adequate power dissipation, special width of power lines is established.

The process of automatic LSI topology synthesis is realized using five main procedures: synthesis of topology (PROC1); definition of lower-level typical cells (PROC2); merging of pairs of typical cells (PROC3); synthesis of interconnection topology (PROC4); union of pairs of typical cells (PROC5). The interconnection procedure is shown in Fig. 3.47.

The currently best known SC are Concorde (Seattle Silicon Technologies), GENESIL (Silicon Compilers), GDT (Silicon Design Labs).

A user forms independently, in interactive mode, the base plan of the chip, selects the number and the types of units. The design process is based on hierarchy of elements, defined in source description. In the stage of scheme design the top-down method is used, and so, input description is provided sequentially as a tree (see Fig. 3.48 for an example). A user can correct and modify the tree obtained in interactive mode.

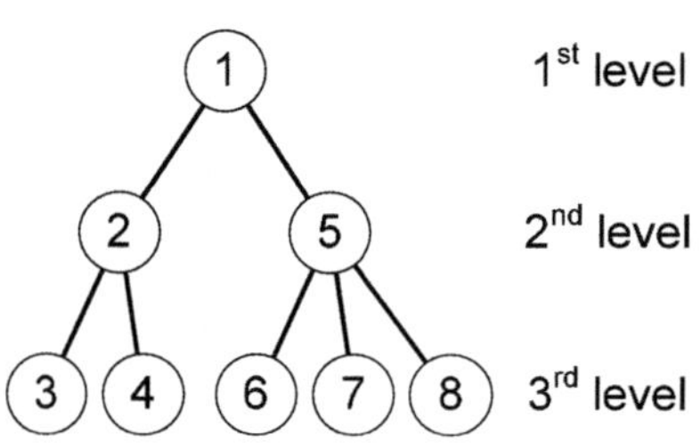

Fig. 3.48. Presentation tree of the device designed (1-8: topological cells)

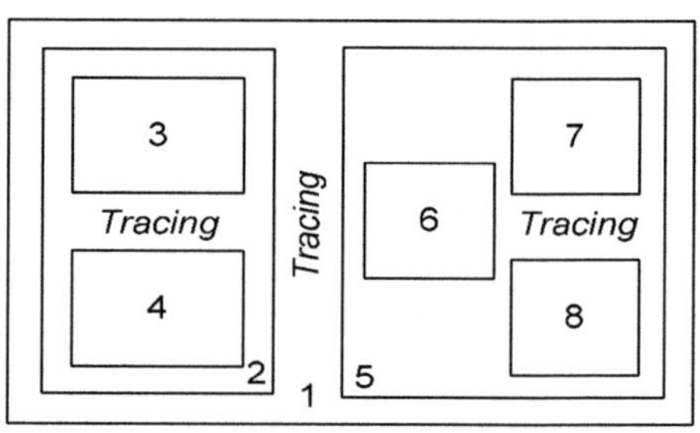

Fig. 3.49. Topological synthesis process for the example of Fig. 3.48

Using information obtained, the topological synthesis of device is made with the bottom-upwards design method. In this process, units of lower level of hierarchy are either in a library, or are generated by procedures using silicon assembler programs. Then, on the basis of level 3 the fragments of hierarchical level 2 are formed by the allocation and routing problem solution blocks, and then the entire device is formed. The topological synthesis process is shown in Fig. 3.49.

It is expected that SC will improve along three main directions:

1. Use of artificial intelligence in CAD systems or the so-called production systems.
2. Comprehensive integration of the synthesis process.
3. System design on higher levels of hierarchy of electronic devices.

Note that in the existing SC the allocation and routing processes are performed automatically, but users can interfere in them and edit intermediate solutions. With the increase of the degree of integrity and density, procedural programming appears as a promising and economically efficient methodology for custom VLSI creation.

Use of SC in VLSI creation increases chip area, but allow the integrity level to reach millions of transistors on one chip.

3.14 Conclusion

This chapter was devoted to questions of construction of the LSI and VLSI CAD systems. Different mathematical models of LSI components were shown and analyzed. Allocation of fragments of topology, combined striping of topology, and control of topology during automated production of photomaps are investigated. Attention is paid to main concepts of construction of LSI and VLSI automated design subsystems, while some ideas as to improvement of VLSI CAND by using hardware-supported software and KB are suggested. This chapter provides the basis for developing new VLSI CAD and selecting the most suitable systems from among the existing ones for definite development purposes.

Design of topology is the most important stage in design of LSI and VLSI. Use of hardware support and silicon compilers in solving engineering problems is a promising method of increasing speed of design speed.

Chapter 4
Development of Genetic Algorithms for Finding Optimal Magnetic Head Parameters and Characteristics

Genetic algorithms (GA) became a powerful instrument for finding optimal solutions in scientific, technical and production problems.

Genetic algorithms differ from other methods of optimization and search [88, 109]. Genetic algorithms:

- analyze and transform the encoded set of input parameters;
- implement search through a population or a set of populations (set of alternative solutions), not only just one solution;
- use target function (appropriateness or fitness function) values rather than its various increments in evaluating alternative solution quality;
- use deterministic, probabilistic and combined rules in the analysis of optimization problems.

In GA the input parameters are first analyzed and a set of alternative solutions called population is generated. Each solution is encoded as a finite sequence in some alphabet. GA works until solution of desired quality is obtained or premature convergence occurs to a local optimum. The process of evolution is based on the analysis of current population of solutions. These current 'parent' solutions generate 'descendants' by random, direct or combined transformations. Then, quality of each alternative solution is evaluated and selection carried out. All evolution models use the principle of 'survival of the fittest' principle, i.e. the least fit solutions are removed and best solutions pass over to the next generation. Then the process is repeated.
Genetic algorithms manipulate populations, composed of 'chromosomes' with mechanisms taken from natural evolution. GA are formally defined as:

$$GA = (Pio,N,PiT,k,T,Lj,A, (TF,LIM,BC),GO,t)$$

where P_i^o is the initial population of alternative solutions (chromosomes), $P_i^o = (P_i^o{}_1, P_i^o{}_2 \ldots P_i^o{}_n)$, $P_i^o{}_1 \in P_i^o$ is a chromosome (alternative solution), belonging to i[th] initial population; n is the number of chromosomes in a population, $N = |P_i^T|$;

V.M. Kureychik et al.: Genetic Algorithms for Applied CAD Problems, SCI 212, pp. 193–210.
springerlink.com

$P_i{}^T{}_k \in P_i{}^T$ is k^{th} chromosome belonging to i^{th} population of T evolution generation; $T = 0, 1, 2,...$ is the sequential number of generations passed by population during evolution; sometimes index of generation is linked with genetic algorithm generation number G; L_j is the length of ith chromosome (alternative solution), i.e. number of genes (elements of encoded solution in a given alphabet), $|P_i{}^T| = L_j$; A is an arbitrary abstract alphabet used for chromosome encoding, for example, $A_1 = \{0,1\}$, $A_2 = \{0,1,2,...,10\}$, $A_3 = \{0,1,2,*\}$, $A_4 = \{A,B,C,D\}$, here * is a label, that is - any symbol from alphabet A_2; *(TF,LIM,BC)* — target function, limitations and boundary conditions defined by a given model of the source problem; *GO* are genetic operators, *t* is the stopping rule.

GA have the following advantages over other optimization methods:

1. simplicity of the object data processing structure, clear structure of the algorithm, defined by the object-oriented approach, at the base of GA;
1. common structure of the data and the algorithm for different purposes, differing only by content (chromosome structure and different genetic operators); this GA property makes of it a universal optimization algorithm for finding optimal solutions;
2. GA customization for various specific problems;
3. possibility of obtaining alternative solutions to a single problem with the same quality but different parameter values.

Only direct search optimization algorithms can yield better results, as they have a well-defined target and the way to reach it in a minimum number of iterations [110].

For finding optimal characteristics and parameters of magnetic heads GA are most suitable because in this case the way towards the optimal solution is unknown, only the target function is defined.

4.1 Genetic Algorithm with Multiple Genotype

4.1.1 Finding Optimal Parameters of Magnetic Heads

It became obvious that finding optimal parameters and characteristics of magnetic heads (MH) requires an algorithm for finding parameters of mathematical models in multi-dimensional solution (parameter) space. An important aspect is that all parameters of mathematical models are initially limited. Besides, the search for optimal solutions of different mathematical models should often be done for their whole, because, for example, the synthesis of glassy dielectric with defined qualities should be done simultaneously with selection of a ferrite so as to match their qualities at the binding. For finding solutions to such problems it is necessary to create a genetic algorithm with multiple genotype (GAMG).

4.1.2 Presentation of Genetic Material

The GAMG is shown here on the examples of optimization of extraneous field, the best output signal of thin-film asymmetric MH, and the highest operation efficiency of a planar single-turn thin-film magneto-resistive head (PTMRH).

The expressions that define the required MH qualities are mathematical formulas, with each parameter value defined inside a specific range. The result is the required characterization of MH. So, MH parameters are included in the chromosome as genes. Each gene changes inside its value range. Every gene of the solution chromosome neatly corresponds to its variation interval and range, defined at the beginning of work. Let us define the solution genotype for the problems mentioned [111].

Problem 1. Determination of the optimal distribution of extraneous field for the asymmetric thin-film MH (see expression (4.1)):

g - front gap size, p - head tip length, Δp - overall length of auxiliary thin-film layers, k1 - coefficient, depending on curvature of field lines of the horizontal component of extraneous field close to front gap and head tip, k_2 - coefficient, depending on gradient of horizontal component of extraneous field, k3 - coefficient depending on gradient of horizontal component of extraneous field close to head tip with auxiliary layers, Hg - extraneous field density in the middle of the gap (x = 0, y = 0).

For problem 1 the chromosome will look like in Fig. 4.1.

g	p	Δp	k_1	k_2	k_3	H_g

Fig. 4.1. A structure of the GAMG chromosome for problem 1

Problem 2. Determination of the best output signal for the asymmetric thin-film MH (see expression (4.2)).

v — data carrier velocity;

M_r — residual magnetization;

w — track width;

δ — thickness of data carrier magnetic coating;

d — distance between head and tape.

Other genes are parameters defining extraneous field. Hence, the genotype for problem 2 is defined by the chromosome structure shown in Fig. 4.2.

g	p	Δp	k_1	k_2	k_3	H_g	v	M_r	w	δ	d

Fig. 4.2. A structure of the GAMG chromosome for problem 2

Problem 3. Determination of highest operating efficiency of PTMRH (see expression (4.4)).

L — core width;

σ — conductivity of intragap layer material (variation interval for this gene is based on consideration that material composition of intragap layer varies in the limited range, i.e. for the best solution another material can be selected);

W — width of magneto-resistive element (MRE);

S — distance from the head working surface to the centre of MRE;

l — depth of core front gap;

μ_1 and μ_2 — relative magnetic conductivity of core and MRE materials;

p — thickness of core poles;

t — MRE thickness;

g_1 — width of core front gap;

g_2 — distance from MRE to opposite core branch.

For this order of genes, the chromosome may look like in Fig. 4.3.

L	W	S	l	p	t	ω	σ	μ_1	μ_2	g_1	g_2

Fig. 4.3. A structure of the GAMG chromosome for problem 3

4.1.3 The Solution Encoding Methods

Correct selection of the chromosome encoding method is an important aspect in development of genetic algorithms [112, 113]. It is convenient to present the chromosome for GAMG as a sequence of integers, each of them defining the value of a multiplier corresponding to change with respect to the initial position:

$$M = F_p / s, \tag{4.5}$$

where M is the gene value, F_p is mathematical model parameter value, s is the parameter variation step defined at the beginning of work.

For example, for problem 3 the chromosome may look as in Fig 4.4:

1	5	69	0	6	5	0	0	28	0	12	5

Fig. 4.4. An encoded solution chromosome

4.1.4 The Target Function

For the problem of highest efficiency of PTMRH (problem 3) the target function (TF) is the average PTMRH efficiency:

$$F(H) = \alpha = \frac{1}{W}\int_0^W \alpha_d(x)dx = \frac{2}{WD}\left\{(k_1 + k_2)\text{sh}\left[k_1\left(l - S - \frac{W}{2}\right) + k_2 W\right] - \right.$$

$$\left. -(k_1 - k_2)\text{sh}\left[k_1\left(l - S - \frac{W}{2}\right) - k_2 W\right] - 2k_2 \text{sh} k_1\left(l - S - \frac{W}{2}\right)\right\},$$

where α_d is differential head performance, W is width of the magneto-resistive element (MRE), k1 is a coefficient depending on field lines curvature close to front gap and head tip l, k_2 is a coefficient, depending on extraneous field horizontal component gradient along OY axis, S is distance between head surface and the middle of MRE.

In this model, α is minimized:

$$F(H) \to \min . \tag{4.6}$$

The TF value for the chromosome is:

$$F(H) = 1.129586741519.$$

4.1.5 Genetic Operators and Algorithm Structure

Crossover and mutation operators are the most important genetic operators used in GAMG [114, 115]. As we use homologous numeric encoding of the solution, it is possible to use standard one-point, two-point or multi-point crossover [116, 117]. Experimental studies, to be presented in Chapter 5, show that the multi-point crossover gives better results. Let us examine the work of this crossover operator on an example.

We apply the multi-point crossover operation to chromosomes H_1 and H_2 (crossover points at genes 1, 4, 5, 7, 10, 11), see Fig. 4.5.

H_1

1	5	69	0	6	5	0	0	28	0	12	5

$F(H_1) = 1.129586741519$

H_2

2	49	1	5	10	3	2	18	11	0	1	3

$F(H_2) = 2.550824569127$

Fig. 4.5. Parent chromosomes for multi-point crossover operator

After crossover operation we obtain two descendants, H_3 and H_4 (Fig. 4.6).

H3

2	5	69	5	10	5	2	0	28	0	1	5

$F(H_3) = 1.007192436231$

H4

1	49	1	0	6	3	0	18	11	0	12	3

$F(H_4) = 2.769351104632$

Fig. 4.6. Descendant chromosomes obtained

From the crossover the descendant H_3 was obtained with better fitness than its parents. So, this type of crossover can significantly improve solutions and can be used in solving the problem of optimal MH characteristics.

Selection of pairs of chromosomes for the crossover operation can be done with elite or random strategy [118]. Experimental studies showed that the elite strategy gives better results of the two.

In terms of the mutation operator, simple, point or exchange mutation can be used in the algorithm [119]. The disadvantage of simple mutation is that the number of changes it causes in a chromosome does not depend on the size of chromosome [120]. Point mutation allows for changing several genes at a time, modifying the chromosome to a greater extent. Exchange mutation can change parameter value to the opposite, which can be useful in the search for solution. Hence, it is convenient to use two mutation operators: point and exchange mutation.

Let us examine the functioning of the point mutation operator. Let a chromosome like in Fig. 4.7 be given, and assume that the mutation operator has selected only the fifth gene. Then, the effect of the mutation operator is as in Fig. 4.8.

H_1

2	5	69	5	10	5	2	0	28	0	1	5

$F(H_1) = 1.007192436231$

Fig. 4.7. A chromosome before application of the point mutation operator

H_1'

2	5	69	5	2	5	2	0	28	0	1	5

$F(H_1') = 1.152586465619$

Fig. 4.8. The chromosome of Fig. 4.7 after application of the point mutation operator

The choice of the selection operator depends on the type of genetic algorithm used. GAMG is a stable state genetic algorithm. Therefore, when it is necessary to add four chromosomes to a population after having used crossover and mutation operators, the selection operator has to remove four chromosomes. As it is necessary to remove duplicate solutions from the population, selection operator will consist of two stages. First, it finds four duplicate chromosomes and removes them. If less than four chromosomes have been removed, it will look through other chromosomes in the order of increasing fitness. Current solution will be removed with probability P_s.

The structural chart of GAMG is shown in Fig. 4.9. This genetic algorithm includes operations accounting for the specificity of the problem solved. Satisfaction of constraints is checked after the use of the genetic operators to prevent appearance of 'illegal' solutions.

4.1.6 Theoretical Evaluation of the Algorithm

For GAMG, time and space complexities are defined as functions of N, where N is the number of mathematical model parameters. Space complexity in the worst case [121] is equal to the biggest of capacity sums of all registers, which were accessed. Memory space is needed for storing chromosome population, and for this nL memory cells are needed, where n is population size (number of chromosomes in a population), L is chromosome length (for GAMG $L=N$). For storing the descendants generated $4L$ memory cells are required. Then, for storing

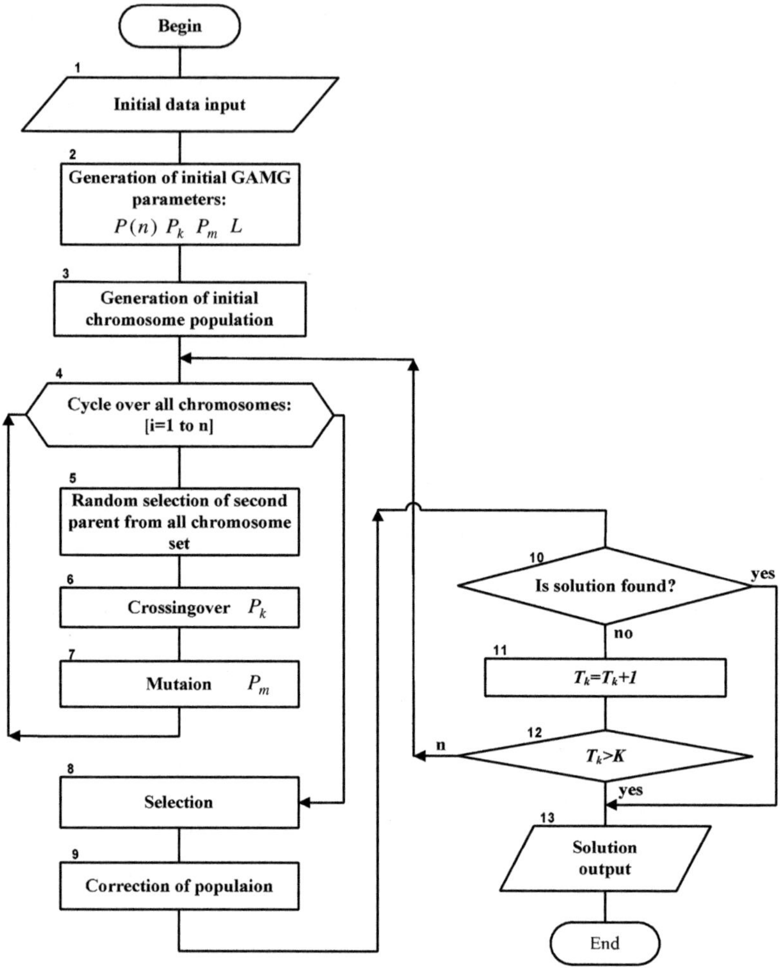

Fig. 4.9. The structural chart of GAMG

population, in the worst case $(n+4)L$ memory cells are required. In this way the space complexity of the algorithm will be:

$$O((n+4)L) . \tag{4.7}$$

Therefore, space complexity can be considered as proportional to *O* (*N*).

Time complexity of a genetic algorithm consists of time complexities of separate procedures being parts of the algorithm [122]: selection, crossover, mutation and chromosome decoding. Time complexity of selection in the worst case is $O((n+4)\log_2(n))$, of crossover - *O* (*L*), of mutation - *O* (*L*), and of chromosome decoding - *O* (*L*). During one generation one crossover operation, two mutation and four decoding operations are made. We must also consider that for the initial population generation each chromosome requires L operations for gene definition and L operations for chromosome decoding. After generating the initial population we must sort population in the order of decreasing TF value, which requires $n\log_2 n$ operations.

Therefore, time complexity of this genetic algorithm equals

$$O(n(2L+\log_2 n)+(3L+n+2\log_2 n+4L)K), \tag{4.8}$$

where K is the number of iterations of the algorithm (of generations).

From the expression (4.8) we can conclude that time complexity of GAMG is proportional to $O(N)$.

After estimating space and time complexities of GAMG, (4.7), (4.8), we can say that for fixed population size n and number of generations K the algorithm has linear theoretical space and time complexities and therefore is of practical value.

4.2 Dynamic Genetic Algorithm

The development of automated design reached the stage on which new ideas, principles and algorithms are needed. The existing optimization algorithms are based on interaction with the end user and must be controlled during the search process and when analyzing the obtained solution in completion stage [123]. Most of algorithms can, namely, miss the proper solution or get stuck in a local optimum, and only the user of the algorithm is able to direct search in the desired direction by inputting new parameters or heuristics. Situation is complicated by the fact that modern optimization algorithms mainly solve problems of polynomial complexity while new technical problems often contain problems of exponential or factorial complexity. For example, problem of synthesis of material with required qualities has exponential complexity [124, 125]. For solving such problems standard optimization algorithms are ineffective. In this situation flexible, self-adjusting optimization algorithms are required [126].

4.2.1 The Purpose of Development of Dynamic Algorithm

First, let us emphasize that selection of materials with desired qualities is one of important stages of DSD MH design process. Because materials used for MH production (ferrite, glass, titanium) have different mechanical strengths, wearing capacities, linear thermal expansion coefficients and other qualities, this stage is most important for junction forming. It must be noted that connection of materials in a junction constitutes the strongest multifunctional structure being the basis of MH. Obviously, automated selection of materials with coordinated qualities allows for a significant improvement of quality of magnetic recording device (MRD) MH.

Let us examine the problem of selection of material with desired qualities on the example of glassy dielectric synthesis for junctions with titanium. The problem boils down to the approximation of elementary $f_k(x_i)$ functions of undefined form.

To solve this problem we use a self-adjusting, dynamic genetic algorithm.

4.2.2 The Structure of Presentation of Genetic Material

The structure of genetic material presentation for dynamic genetic algorithm (DGA) of formula selection is based on genetic algorithm for formula approximation (GAFA). In GAFA a chromosome is presented as a set of mathematical functions: one-parameter (sin, cos, tg, exp, etc.), two-parameter (+, –, ×, /, etc.) and three-parameter (if- then- else-), composing a formula $f_k(x_i)$ that approximates the diagram of experimental values [123]. The argument of these formulas is x_i (in our case: vector of percentage shares of material components in composition obtained). Then, a chromosome (alternative solution) may look as shown in Fig. 4.10 [124].

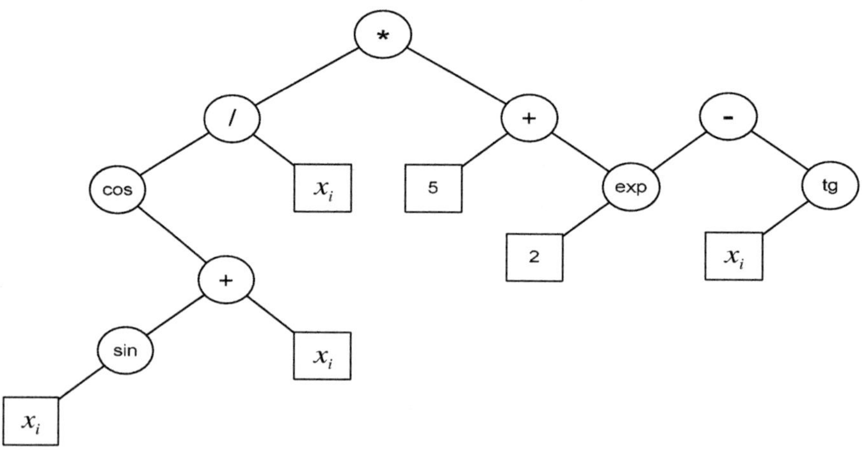

Fig. 4.10. Tree structure of a GAFA chromosome

The chromosome from Fig. 4.10 corresponds to the function:

$$f_k(x_i) = \left[\frac{\cos(x_i + \sin(x_i))}{x_i} \right] \cdot [5 + \exp(2) - tg(x_i)] \cdot$$

4.2.3 The Method of Encoding Solution

For encoding chromosomes in DGA we use the structure like in Fig. 4.11.

+ ¦ 0	sin ¦ 0	cos ¦ 0	/ ¦ 0	* ¦ 2	5 ¦ 1	+ ¦ 0	exp ¦ 0	2 ¦ 1	- ¦ 2	tg ¦ 0

Fig. 4.11. The DGA chromosome encoding scheme

This structure consists of two-symbol genes. The first symbol is a mathematical operation from the list of formulae. Each operation is encoded by a unique code in the chromosome, and a longer list of formulas used gives the algorithm broader approximation capabilities. The second symbol is the gene decoding sign that can take three values:

0: in this case, if expression is completed, this gene is an elementary function, whose argument is an expression accumulated by the moment of this gene processing; if expression is incomplete or it is the first gene, then;

1: the gene includes an integer;

2: the gene includes an elementary function, whose argument is an expression from non-processed genes to the right.

In GAFA, chromosomes may have different length, as a junction in the chromosome tree structure is a crossing point. Variable length allows for the use mathematical formulae of any length to obtain accurate formula approximations. Note that mathematical operation or variable can be replaced by mathematical expression and can be removed. Changing the chromosome length does not interrupt the logic of GAFA, because for two crossing chromosomes the initial point(s) of modification is set randomly. After modification or crossing the chromosome is checked for correctness.

4.2.4 Target Function

Target function for DGA chromosomes is average approximation error:

$$Fitness[l] = \left| f_k(x_i) - y_j \right|;\ \overline{i} = 1{,}9;\ \overline{k} = 1{,}27;\ \overline{j} = 1{,}3;$$

where $Fitness[l]$ is TF value for chromosome l; $f_k(x_i)$ is value of kth approximated function (x_i — vectors of experimental glasses 1-10 components percentage values); y_j — vectors of qualities of the glasses: linear thermal expansion coefficient (LTEC), T_g, H.

4.2.5 Genetic Operators and Algorithm Structure

The gist of GAFA lies in modification and crossing of homological parts of parent chromosomes using GA operators. As a result we have descendant chromosomes with changed mathematical formula.

Let us look at an example of using one-point crossover operator. Parent chromosomes H_1 and H_2 are shown in Fig. 4.12.

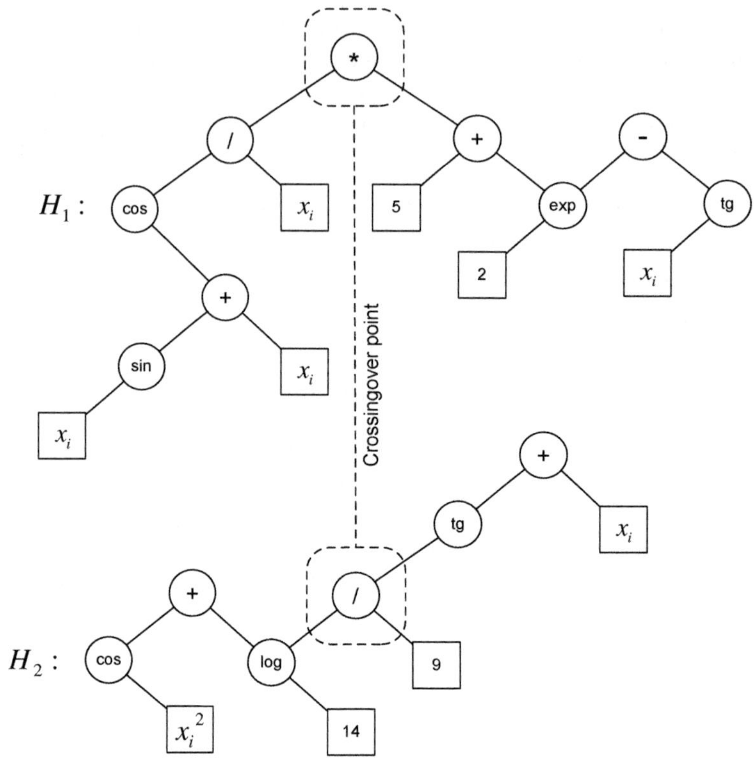

Fig. 4.12. Parent chromosomes for an example of one-point crossover

As a result of crossover operation used we get descendants H_3 and H_4, shown in Fig. 4.13. Decoding of the descendants H_3 and H_4 gives the following results:

$$H_3 : f_k(x_i) = tg\left[\frac{\cos\left(\dfrac{\sin(x_i) + x_i}{x_i}\right)}{9}\right] + x_i ,$$

$$H_4 : f_k(x_i) = \left[\cos\left(x_i^2\right) + \log(14)\right] \cdot [5 + \exp(2) - tg(x_i)]$$

It was concluded from multi-parameter problem solving and from investigations of classical algorithms [127—131] that the desired algorithm should not contain constant parameters. Thus, the algorithm decides by itself which parameters should be changed if during certain number of generations the target function does not improve. This decision is due to the fact that the genes of this problem are

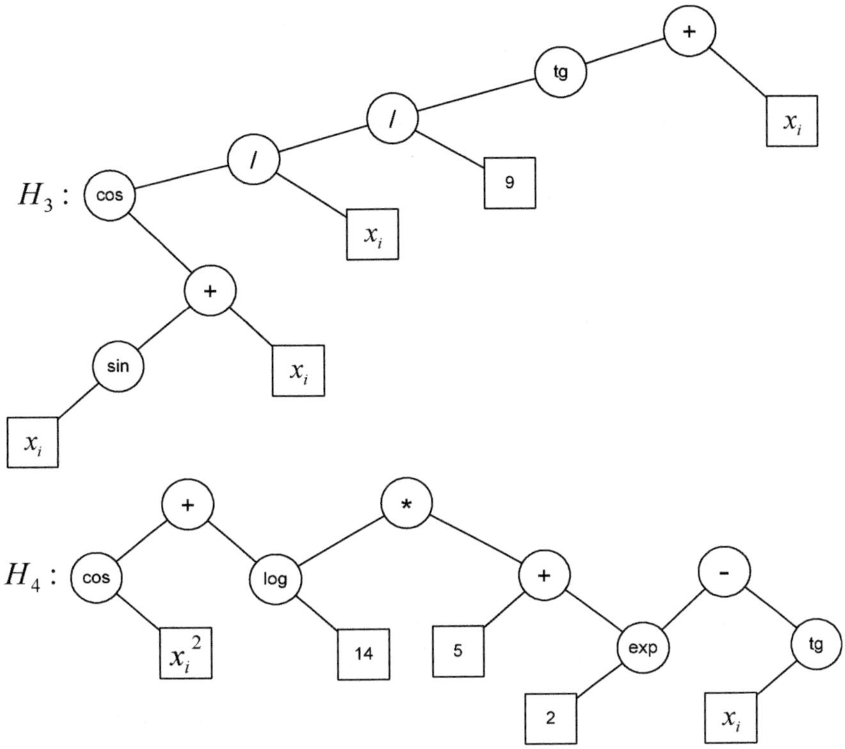

Fig. 4.13. Descendants obtained as a result of application of one-point crossover operator

integer, not binary [132], so that problem dimensions increase exponentially. Further, the length of a chromosome (alternative solution) should be temporary, the algorithm itself regulating and controlling this length. Hence, the algorithm should be self-adjusting, leading to fast and high-quality problem solution. The suggested algorithm is called dynamic genetic algorithm (DGA). It is based on the concept of genetic algorithm with dynamic operators [133].

The basic idea of DGA is modeling of natural evolution processes in animate nature [28]. The main features of the algorithm [134] are:

1. Each object in DGA (chromosome, genetic operator, population) has a lifetime parameter, measured in number of iterations. 'Old' DGA objects do not disappear after generation of new population and can be used in subsequent iterations. The purpose of this feature is to prevent the loss of collected genetic material as it can be useful in improving the solution.
2. DGA operators contain an efficiency estimation parameter. It allows for selecting the operators that are able to improve TF or for removing ineffective operators. This parameter is used as in DGA the genetic operators can change, too, like in animate nature not only living species evolve, but environmental

conditions change as well. The evolution of DGA operators is done recursively, i.e. operators are modified by themselves. The suggested script language for describing operator actions is similar to a programming language. DGA operators form populations and evolve. The TF of an operator is the number of iterations during which this operator will improve TF of the formula chromosome population. Chromosome genes are operator strings of executed actions. The initial set of DGA operators consists of known genetic operators: 1) one-point, two-point and multi-point crossover operators (CO); 2) point and multi-point mutations (MO), deletion, insertion, translocation and other mutation operators; 3) different reproduction and selection operators [117], and other ones.

3. Possibilistic parameters of DGA, the crossover, P_k, and mutation, P_m, probabilities are also adjusted by the algorithm during the work. Also, the efficiency parameter determined by number of iterations needed for DGA objects improvement is set here.
4. Length of a chromosome is regulated by its TF value, and so, if a chromosome is too 'long' or too 'short', it will not provide an accurate solution and will eventually die.
5. Population size is regulated by the efficiency coefficient based on the number of chromosomes that improve current TF value.

With respect to random generation of initial chromosome population, initial population of DGA operators, initial values of the crossover, P_k, and mutation, P_m, probabilities, and population size, DGA have built-in database to store solutions and the algorithm settings, which allowed for getting the optimal solution [135]. This database is used in the DGA self-learning unit. In the initial stage this unit accesses the solution database to match initial input data and the stored variants of solved problems.

If close correspondence is established for the entire set of initial data set or its part, the self-learning unit loads settings stored in database to improve the speed of finding optimal solution.

The structural chart of DGA is shown in Fig. 4.14.

Let us consider an application of pattern theory [28, 136] to binary chromosomes [137] in DGA. A pattern H is defined by elements of the set $\{0,1,*\}$: $H \in \{0,1,*\}$. DGA can be described using variables: $P(n)$ - chromosome population; n - population size; $N(H,t)$ - number of copies of pattern H in generation number t; M - set of population chromosome numbers, in which there is pattern H; $D(H)$ - length of pattern H; $L(H)$ - length of string containing template H; f_i - target function of chromosome i;

If the probability of selecting a descendant through random selection is

$$P_{sel} = (f_{\max} - f_k) / \sum_{i=1}^{n} (f_{\max} - f_i),$$

then the average number of patterns H in the population, N_s, is

$$N_s = n \sum_{j \in M} (f_{\max} - f_j) / \sum_{i=1}^{n} (f_{\max} - f_i).$$

It is obvious that $F_s(H) = \sum_{j \in M} (f_{\max} - f_j) / N(H,t)$,

$$F_p = \sum_{i=1}^{n} (f_{\max} - f_i).$$

Here, $F_s(H)$ is the average utility of descendants containing pattern H; and F_p is average utility of all descendants of the population. So,

$$N_s = N(H,t) F_s(H) / F_p.$$

For one-point crossover, occurring with probability P_k, the pattern enters the descendant chromosomes with probability

$$P_{sk} = (1 - P_k D(H)/(l-1)).$$

During mutation, occurring with probability P_m for each position, the probability of preserving the pattern is $P_{sm} = (1 - P_m L(H))$.

As each chromosome in DGA has the lifetime parameter t_h (number of lifetime iterations of a chromosome), the probability of pattern being alive is

$$P_{st} = \left(\frac{K}{t_h - k_h} \right).$$

Consequently, the probability of survival of pattern H is

$$P_s = (1 - P_k D(H)/(l-1))(1 - P_m L(H)) \left(\frac{K}{t_h - k_h} \right).$$

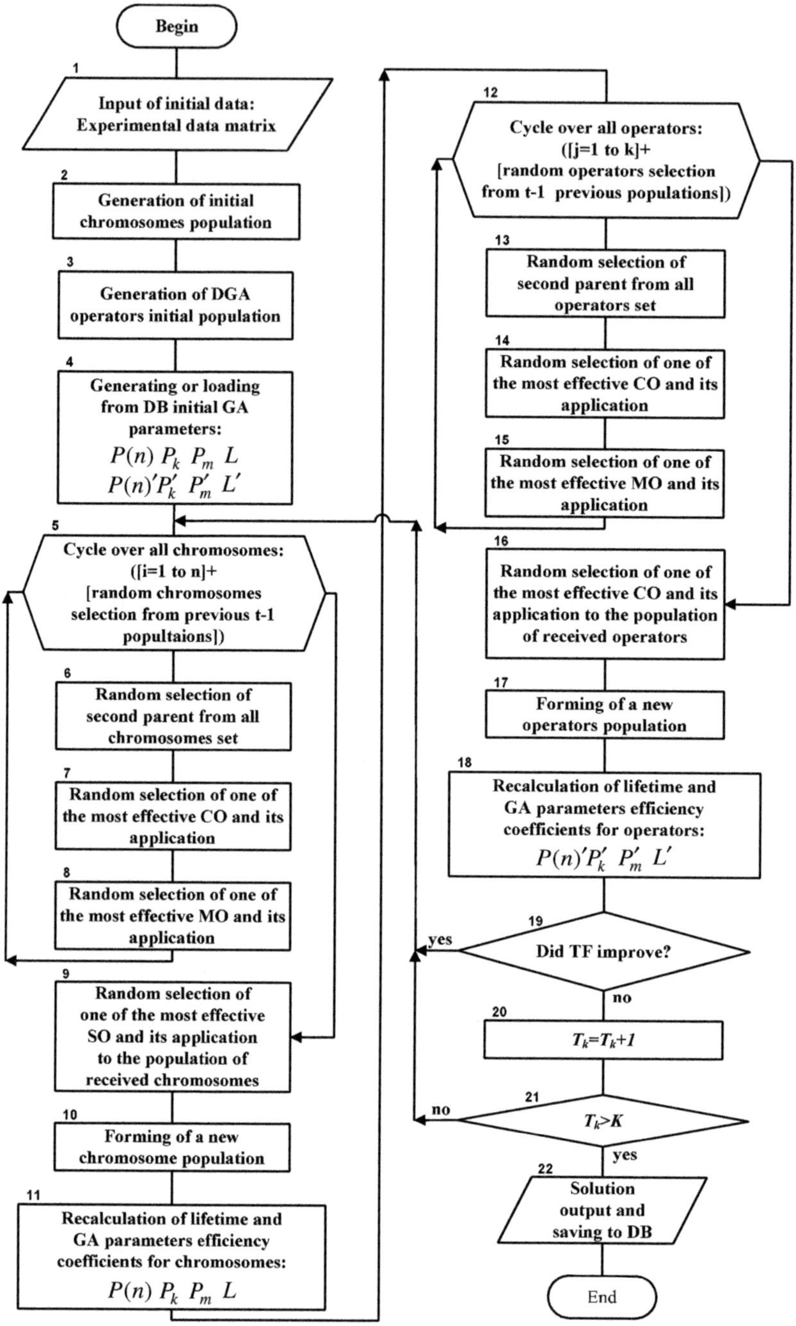

Fig. 4.14. Structural chart of DGA

Consequently, the pattern theory formula for DGA will be as follows:

$$N(H,t+1) = N(H,t)P_s F_s(H) / F_p .$$

This expression defines the number of patterns that will survive at iteration t+1.

At the basis of DGA development there is the idea that GA are fully defined by their structure, with parameters set by the user. It is obvious that the structure of genetic algorithm and its parameters directly influence the quality and speed of solution. Usually, in GA formation the developer follows own intuitive surmises that can be mistaken in one situation and successful in another. Furthermore, GA with a strictly fixed structure are tuned to one problem and are not universal. So, such algorithms can 'pass by' the desired solution, find it through random coincidence of parameter values, or end in a local optimum.

As all the initial DGA parameters and populations can be far away from the optimal ones, in first iterations the TF value changes slightly. However, due to the built-in dynamic elements allowing for in-process change of chromosome length, genetic operators, algorithm parameters and for evaluating efficiency of each element, DGA shall in the final stage reach the optimal solution. It is important that the process of solution search goes on without user participation. The DGA self-learning unit reduces time requirements, allowing subsequent problems to be solved faster by using parameters of previous solutions.

4.2.6 Theoretical Evaluation of the Algorithm

The time complexity of the algorithm is:

$$O(S_p(L+N^2+\log_2 S_p)+(6L+S_p+4\log_2 S_p+8N^2)K),$$

where S_p is average population size; L is average chromosome length; N is the number of mathematical model parameters; K is the number of iterations.

The DGA tunes the values of parameters for each problem solved towards the optimal ones. Therefore, notwithstanding the self-adjusting elements, DGA has polynomial complexity, proportional to $O(N^2)$ and so is of practical value.

4.3 Conclusions

1. Genetic algorithms are presented: a GA with multiple genotypes, GAMG, allowing for finding optimal MH parameters and characteristics, and dynamic genetic algorithm, DGA, used to determine the composition of material with desired qualities.
2. A method of representing genetic material in GAMG is outlined; it allows for preventing of 'illegal' solutions.

3. A method of representing genetic material in DGA is outlined; it allows for rejecting most of the ‘illegal’ solutions. This method is faster than the existing analogues.
4. The target function of DGA is formulated; it allows for evaluating the solutions in approximation of experimental data.
5. The genetic operators and the structural chart of GAMG are provided.
6. The architecture, the structure of genetic operators and the self-adjusting elements of the DGA are presented. The dynamic structure of DGA distinguishes it advantageously from the classical genetic algorithms with constant structure and parameters.
7. The theoretical estimates of space and time complexities of GAMG and DGA are developed.

Chapter 5
Experimental Investigation of Algorithms Developed

5.1 The Purpose of Experimental Investigation

After a CAD problem has been solved, its efficiency should be checked. The problem of finding optimum MH characteristics and parameters is considered in the book in terms of randomly-directed algorithms. The efficiencies of algorithms can be compared on the basis of:

- theoretical investigation, i.e. comparison of time and space complexities of the algorithms;
- experimental investigation, i.e. performing a series of experiments and comparing experimental data for various algorithms.

The objects of investigations in this work are:

- the genetic algorithm with multiple genotype;
- the dynamic genetic algorithm.

The experiments conducted had two purposes:

- investigation of genetic search mechanisms in terms of solving problems;
- investigation of suggested algorithms efficiency.

Investigation of genetic search mechanisms consisted in determination of the influence exerted by the following control parameters: selection type, crossover type, probability of transfer of a gene from one chromosome to another (for multipoint crossover), mutation probability and population size.

Investigation of suggested algorithmic efficiency consisted in determination of space and time complexities and comparison of results obtained from genetic algorithms with the optimal ones, or, if finding of optimal results was impossible, with results of other algorithms. For the algorithms developed we created a software package using C++ language. Experiments were made on a PC AMD Athlon with 650 MHz processor and 256 MB of RAM.

During the experimental investigations of the algorithms developed it is necessary to establish:

- dependence of solution time upon a number of parameters of the mathematical model for GAMG and the average chromosome length for DGA;
- dependency of main memory usage by GAMG and DGA.

These dependencies are estimated using the least squares method [138, 139].

V.M. Kureychik et al.: Genetic Algorithms for Applied CAD Problems, SCI 212, pp. 211–223.
springerlink.com

5.2 Investigation of Genetic Algorithm with Multiple Genotype

5.2.1 Definition of Optimal Parameters

In practical use of genetic algorithms it is important to find the values of control parameters that would provide for high solution quality with the least possible number of generations. In the steady state genetic algorithms these parameters are: selection type, crossover type, probability of gene transfer from one chromosome to another (for multi-point crossover), mutation probability and population size.

First, influence of selection type on quality of solutions was investigated. Elite, random selection and standard one-point, standard two-point and multi-point crossover were examined. Multi-point crossover was used with different values of probability of gene transfer from one chromosome to another, namely 0.1, 0.2, 0.3, 0.4, 0.5. Probabilities higher than 0.5 were not examined, as they would be statistically equivalent to probability of 1.0. For investigating these parameters mutation probability P_m was set at 0.05 and population size n was 50. The number of generations was set at 800. For each variant of genetic parameter settings 50 tests were performed.

In the second stage the influence of population size and mutation probability was analyzed. For this study, selection type, crossover type and gene transfer probability were all set equal to 1, as having given the best results in the previous stage. Population size changed from 50 to 150 with the step of 50, mutation probability changed from 0.05 to 0.20 with the step of 0.05. The number of generations was set at 800.

In investigation of GAMG the test problem selected had 12 parameters of the mathematical model.

The results of the first stage are provided in Table 5.1. The columns ‘Crossover type’ and ‘Selection type’ show the types of crossover and selection used in tests. Column ‘P_k’ shows the probability of gene transfer from one chromosome to another. In this column the sign ‘–’ (for one-point and two-point crossover) means that this probability is not used. During the test, generation with target function value of 1 was determined. Minimum, maximum and average numbers of generations for test series are provided in the table.

Experimental investigations showed that selection type strongly affects solution quality, with elite selection allowing for getting better results, for the problem considered, than random selection. Among the crossover operators the best was multi-point crossover, the second - two-point, and the worst - the one-point crossover. So, use of multi-point crossover with P_k probability of 0.20 and elite selection gave the best results.

In the second stage it was analyzed how population size and mutation probability affect the quality of solutions from GAMG. Results of experimental investigations are shown in Table 4.2. In ‘Population size’ column respective values for test series are given. The ‘P_m’ column shows gene mutation probability in a chromosome.

Table 5.1. Results of tests in stage 1

Crossover type	Selection type	P_k	Number of generations		
			min	max	average
one-point	elite	–	184	747	475
	random	–	428	765	595
two-point	elite	–	148	647	415
	random	–	361	702	525
multi-point	elite	0.1	202	710	430
		0.2	126	554	308
		0.3	138	766	486
		0.4	148	796	458
		0.5	122	757	474
	random	0.1	404	763	606
		0.2	305	766	533
		0.3	320	796	561
		0.4	349	723	577
		0.5	353	640	493

During the tests generation with target function value of 1 was defined. The minimum, maximum and average values of generation numbers for test series are provided in Table 5.2.

Table 5.2. Results of tests in stage 2

Population size	Pm	Number of generations		
		min	max	average
50	0.05	560	717	621
	0.10	578	670	591
	0.15	293	735	508
	0.20	144	764	415
100	0.05	574	716	614
	0.10	438	713	548
	0.15	427	720	541
	0.20	309	575	412
150	0.05	285	585	395
	0.10	636	750	677
	0.15	409	745	547
	0.20	90	687	364

Experimental investigations showed that best value of mutation probability is 0.20. For this value the best results were obtained in runs with population sizes of 100 and 150. We can also say that increase of GAMG population size allows for a slight improvement of solution quality. It can be seen from Table 5.2 that the best solutions were obtained for population size 150 and mutation probability 0.20.

Considering the results of experimental investigations we can recommend the following parameter settings for the practical use of GAMG:

population size n = 150;
elite selection;
multi-point crossover with probability of gene transfer from one chromosome to another, $P_k = 0.20$;
mutation probability, $P_m = 0.20$.

5.2.2 Space and Time Complexity

In order to determine empirical space and time complexity of GAMG tests were made for chromosome sizes from 3 to 30 with step of 3. The size of each series is defined by the residually large numbers table based on Bernoulli theorem [140], depending on confidence probability and the error allowed for. Confidence probability or 'measure of risk' is defined by the probability with which the respective conclusion is made. The closer the value of examined probability to 1 is, the higher the conclusion qualityis. In the practice of scientific investigations probability $P = 0.95$ is commonly used. Admissible error E is set depending on the character of the event examined. In most cases E is set to 0.05. For these values of probability and admissible error the number of observations will be 352. To reduce machine time we set confidence probability to 0.90 and E to 0.10, and then the sample size is equal to 45.

The results of experiments with GAMG in terms of space and time complexity are given in Table 5.3. In this table column '#' shows the number of tests series, column 'Number of parameters' shows the number of GAMG parameters in test problems. Average time of reaching solution for each series is provided in column 'Solution time, sec.' and the average size of memory used for each series is given in column 'RAM size, KB'.

We obtained from regression analysis the following empirical expressions:

- dependence of time to solution on the number of GAMG parameters:

$$f(x) = 14.904762 \cdot \mathrm{x} - 52.428571, \tag{5.1}$$

- dependence of memory volume used on the number of GAMG parameters:

$$f(x) = 22.761905 \cdot \mathrm{x} + 99.02761, \tag{5.2}$$

As a result of experimental investigations of GAMG we obtained estimates of time to solution and used memory that are linear functions of the number of GAMG parameters. These experimental results are conform to the theoretical ones.

Table 5.3. Examination of time and space complexity of GAMG

#	Number of parameters	Solution time, sec.	RAM size, KB
1	3	6.107015	165.0673
2	6	37.00327	237.007
3	9	83.33751	310.6247
4	12	123.8945	354.5536
5	15	162.721	440.5316
6	18	219.0505	511.4094
7	21	270.4563	606.5914
8	24	314.0995	662.3664
9	27	350.8192	714.0846
10	30	408.8172	772.6448

5.2.3 Comparative Characterization

GAMG was implemented for an IBM PC compatible computer running Windows 98-XP operating systems.

For purposes of comparison the test problem from subparagraph 4.1.2 of problem 3 was used. The comparison of space and time characteristics was carried out for the following algorithms: branch-and-bound, simulated annealing and GAMG. The results are provided in Table 5.4.

In Table 5.4, columns 'Solution time, sec.' and 'RAM size, KB' show average time to solution and memory size for the series of 50 runs.

On the basis of data from Table 5.4 we can say that GAMG is the fastest of all algorithms investigated, but has bigger memory requirements than the simulated annealing algorithm.

Table 5.4. Comparison of selected algorithms in terms of time and space characteristics

#	Compared algorithms	Solution time, sec.	RAM size, KB
1	Branch-and-bound algorithm	1563.2945	749.3917
2	Simulated annealing algorithm	958.3371	312.5127
3	Genetic algorithm with multiple genotype	123.8945	354.5536

5.3 Investigation of Dynamic Genetic Algorithm

5.3.1 Selection of Algorithm Parameters

Unlike GAMG, for which it is necessary to choose best parameters for purposes of empirical investigation, the self-adjusting DGA procedures choose optimal parameters during execution of the algorithm. Therefore, experimental investigation of DGA consisted in examination of changes in the parameters of the algorithm.

This analysis was performed during solving of the problem of designing the material with desired qualities on the example of synthesis of glassy dielectrics for junctions of glass with titanium. The chosen test duration was 2000 iterations. Most important DGA parameters were investigated:

P_k — crossover probability;

P_m — mutation probability;

n — population size;

L_{av} – average length of the chromosome in a population.

For each stage of analysis of parameter change the series of 50 experiments were made.

In the first stage of investigation we examined gross variation of crossover and mutation probabilities. The results are contained in Table 5.5. In this table the 'iteration' row shows iteration numbers from 0 to 2000 with the step of 125. 'P_k' and 'P_m' show the corresponding values of crossover and mutation probabilities of the algorithm.

Table 5.5. Changes of the DGA crossover and mutation probabilities

Iteration	**0**	**125**	**250**	**375**	**500**	**625**	**750**	**875**	**1000**
P_k	0.2	0.4	0.36	0.543	0.56	0.56	0.56	0.56	0.551
P_m	0.43	0.43	0.33	0.31	0.31	0.31	0.31	0.31	0.31
Iteration	**1125**	**1250**	**1375**	**1500**	**1625**	**1750**	**1875**	**2000**	
P_k	0.46	0.31	0.76	0.8	0.23	0.463	0.43	0.43	
P_m	0.38	0.28	0.14	0.12	0.14	0.16	0.16	0.16	

The diagram of changes in the values of parameters considered is shown in Fig. 5.1. On the basis of this diagram we can say that during initial 400 iterations DGA tried to find the crossover P_k and mutation P_m probabilities that would improve the current solution. Then, from iteration 400 to 900 DGA does not change these parameters as TF improves at every iteration (see Fig. 5.6). Starting with iteration 900, though, the value of TF changes slowly, so during iterations 900 through 1875, the algorithm tries to find new values of the two probabilities. As a result, at around iteration 1875 the DGA establishes new parameter values, $P_k = 0.43$, $P_m = 0.16$, allowing for improving the TF.

The second stage of DGA investigation was examination of changes in population size n with the number of iterations. The results of empirical study of population size changes are shown in Table 5.6. Here, 'Iteration' shows iteration numbers from 0 to 2000 with the step of 125, while 'n' gives population sizes for corresponding iterations of the algorithm.

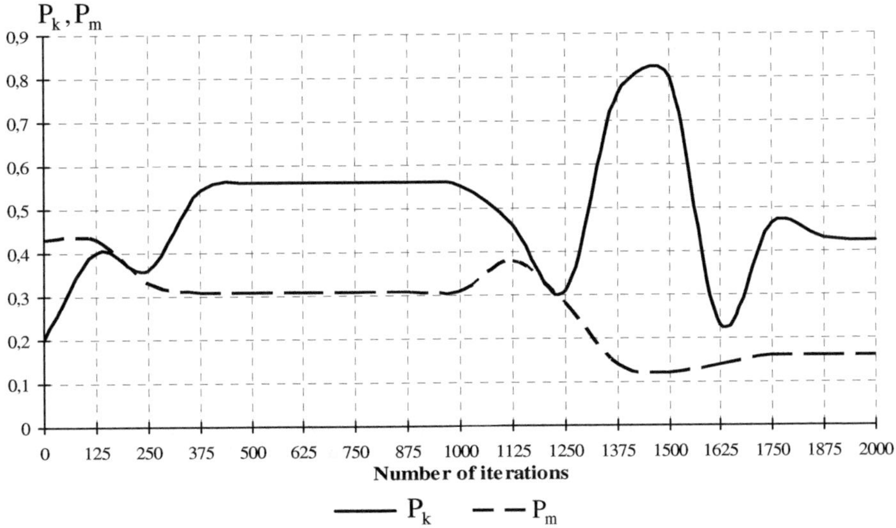

Fig. 5.1. Diagram of changes in DGA crossover and mutation probabilities

Table 5.6. DGA population size changes

Iteration	**0**	**125**	**250**	**375**	**500**	**625**	**750**	**875**	**1000**
n	56	80	102	113	115	115	115	115	115
Iteration	**1125**	**1250**	**1375**	**1500**	**1625**	**1750**	**1875**	**2000**	
n	121	135	149	128	119	112	112	112	

The diagram of Fig. 5.2 shows the changes of parameter *n*. It can be seen that during initial 400 iterations populations size grows from 56 to 115. DGA establishes that increase of population size leads to improved solutions. Between iterations 400 and 1000 the algorithm does not change population size *n* because improvement of TF exceeds the threshold of changing the value of *n*. Then, between iterations 1000 and 1370 the DGA increases population size from 115 to 149, but the value of TF decreases. Hence, from iteration 1370 to 1830 the algorithm changes the value of *n* to 114, this value allowing for continuing TF improvement. Consequently, beyond iteration 1830 the DGA does not change population size.

In the third stage of analysis of DGA functioning we examined changes of average chromosome length in a population, L_{avg}, with the number of iterations. The experimental data on changes of L_{avg} are given in Table 5.7. The row 'L_{avg}' shows the values of average DGA chromosome length, corresponding to appropriate iteration numbers.

The graphical illustration of the changes in L_{avg} is shown in Fig. 5.3.

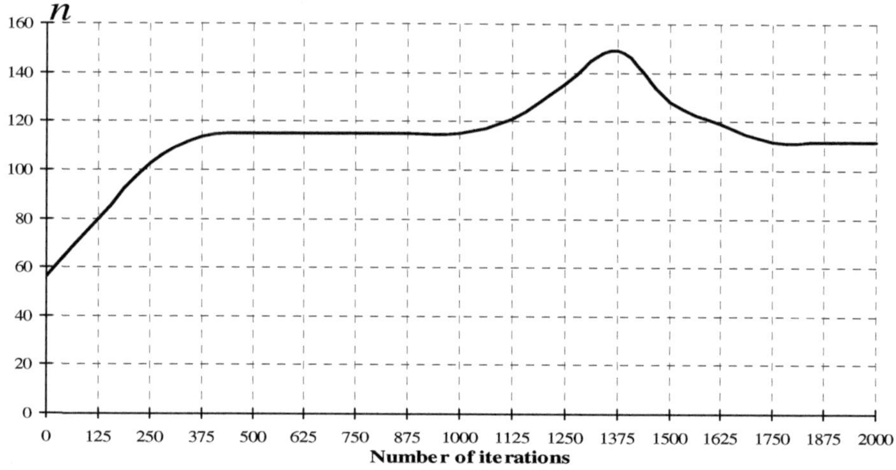

Fig. 5.2. Changes of DGA population size with iterations

Table 5.7. Changes in the average DGA chromosome length

Iteration	**0**	**125**	**250**	**375**	**500**	**625**	**750**	**875**	**1000**
L_{avg}	12	14	17	20.7	21	21	21	21	21
Iteration	**1125**	**1250**	**1375**	**1500**	**1625**	**1750**	**1875**	**2000**	
L_{avg}	20.3	15	18.1	16.2	21.3	22	22	22	

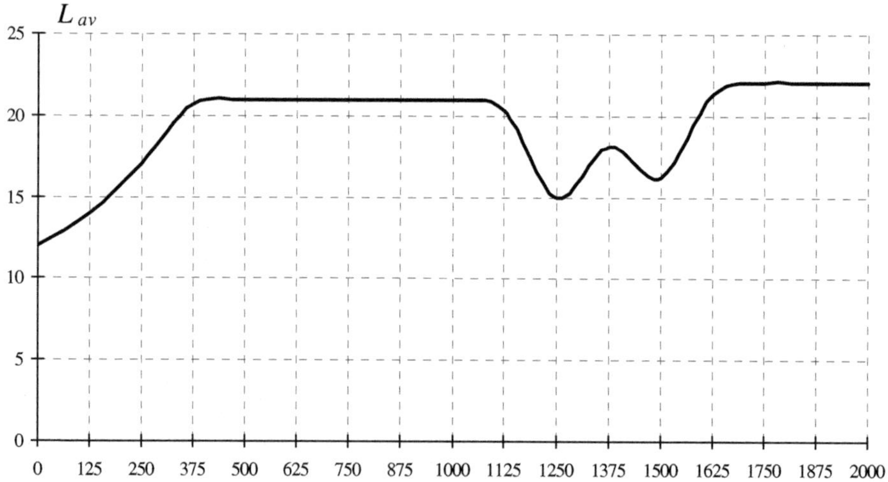

Fig. 5.3. Changes in the average DGA chromosome length with iterations

In the diagram of Fig. 5.3 we can trace the changes in the average chromosome length in the population. During initial 430 iterations the DGA increases the average length of chromosome from 12, characterizing the randomly generated initial chromosome population, to 21. During iterations 430 to 1080 the algorithm does not change the average chromosome length L_{avg}, as solution keeps improving. From iteration 1080 on the TF does not improve, and so DGA tries to pick up a new value. As a result, the algorithm performs a search for the appropriate value of the parameter, by varying it up and down till at around iteration 1650 the average chromosome length stabilizes at 22, allowing for the continued improvement of TF towards the optimum.

An important aspect of DGA is that its self-adjusting unit allows for using different parameter values that improve solution quality during the entire algorithm work. After the experimental investigation of changes in DGA parameters we can say that the algorithm developed reacts correctly to a halt in TF improvement and finds new values of parameters, allowing for improving the solution with brand new parameters combinations. So, unlike classic genetic algorithms [21, 110, 141—142], DGA justifies its name by the parameter change dynamics, aiming at parameter combinations that lead to solution improvement.

5.3.2 Space and Time Complexity

The conditions set for investigating space and time complexity of DGA were the same as for experiments with GAMG, described before. A series of tests were performed by changing the average chromosome length from 12 to 39 with the step of 3. The results of experiments are provided in Table 5.8.

In this table, column '#' shows the number of tests series, 'Average DGA chromosome length' shows average length of DGA chromosome for one series of tests. The remaining notations are like in previous case.

Following empirical expressions were obtained from regression analysis:

- dependence of time to solution on average chromosome length:

$$f(x) = 1.5816 \cdot x^2 - 35.085 \cdot x + 450.615, \tag{5.3}$$

- dependence of memory volume required on average chromosome length:

$$f(x) = 49.375 \cdot x - 226.623. \tag{5.4}$$

Diagrams illustrating these dependencies are shown in Figs. 5.4, 5.5.

On the basis of these figures we can say that the regression models obtained reflect the experimental results sufficiently accurately, as deviation of regression lines from experimental values is small.

Resulting from these experiments we have a polynomial dependence, proportional to $O(N^2)$, of time to solution on the average chromosome length (Fig. 5.4). Experimental estimation of space complexity (Fig. 5.5) showed that in DGA

Table 5.8. Empirical investigation of DGA time and space complexity

#	Average DGA chromosome length	Solution time, sec.	RAM size, KB
1	12	254.5428	409.2886
2	15	284.1416	515.1559
3	18	339.6665	685.1035
4	21	420.7878	785.0721
5	24	547.6489	904.5897
6	27	640.7204	1074.025
7	30	788.5533	1244.11
8	33	947.9477	1391.115
9	36	1176.041	1520.931
10	39	1519.367	1699.544

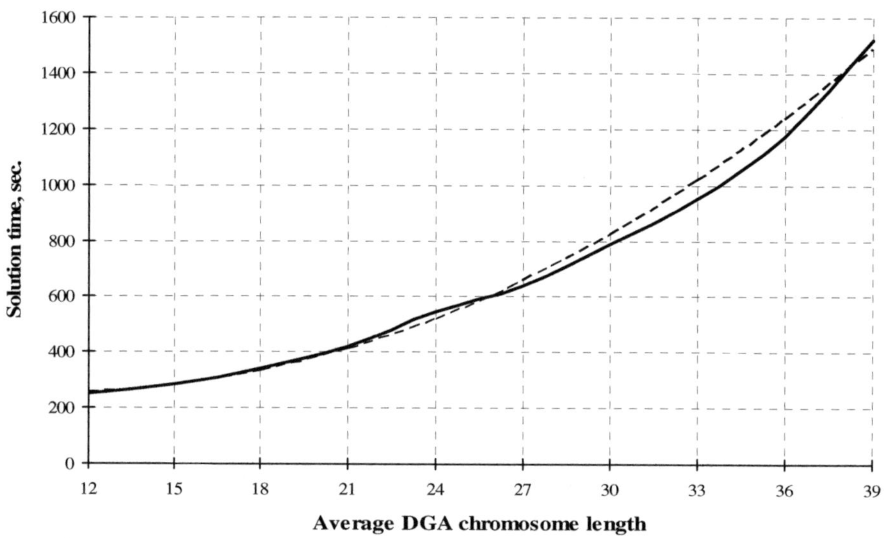

Fig. 5.4. Dependence of time to solution on average DGA chromosome length

the dependence of memory used on the average chromosome length is proportional to $O(N)$.

5.3.3 Comparative Characterization

The DGA was implemented as customizable software for IBM PC compatible computer under Windows 98 XP operating systems. Minimum system requirements are: Pentium 200MHz MMX, 64 MB of RAM, Windows 98. Tests were

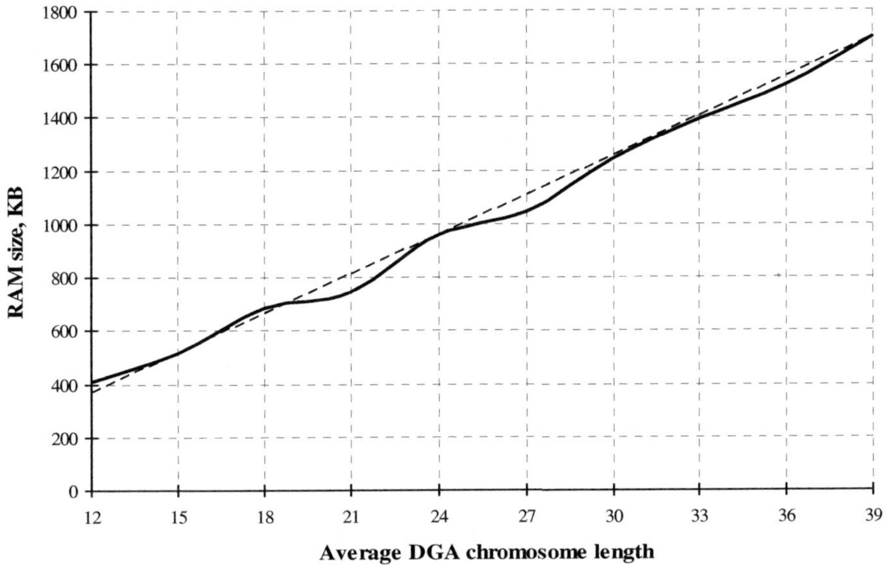

Fig. 5.5. Dependence of memory size used on average chromosome length

made on a PC computer with AMD Athlon 650MHz processor, 256 MB of ram running Windows XP operating system.

To examine the efficiency of DGA we compared it with the classical algorithms. In this investigation we used the test problem of approximation of experimental data table in the problem of selecting the material with desired qualities on the example of synthesis of glassy dielectrics for junctions of glass with titanium. The results of the comparison are presented in Fig. 5.6, where characteristics of speed and quality of solution are shown for the following algorithms:

HA - a heuristic algorithm for formula approximation; HA includes heuristic procedures for defining correspondence of elementary functions to the approximated function; HA performs random search among the possible variants of elementary formulas;

SGA is a simple genetic algorithm [127], a GA including a set of simple genetic operators: one- and two-point crossover operator, point mutation, 'roulette wheel' selection;

AGA is an adaptive genetic algorithm [130], a GA that includes specially selected operators and algorithm parameters that improve solution; AGA selects randomly the initial genetic operators and parameters: when TF improvement stops, algorithm selects other random parameters or genetic operators defined at the beginning of functioning; the degree of success of AGA for a given problem is defined by the quality of selected parameters and operators; this, in turn, depends heavily on the user of the algorithm;

DGA is the proposed dynamic genetic algorithm.

The target function used is the error of approximation of experimental data.

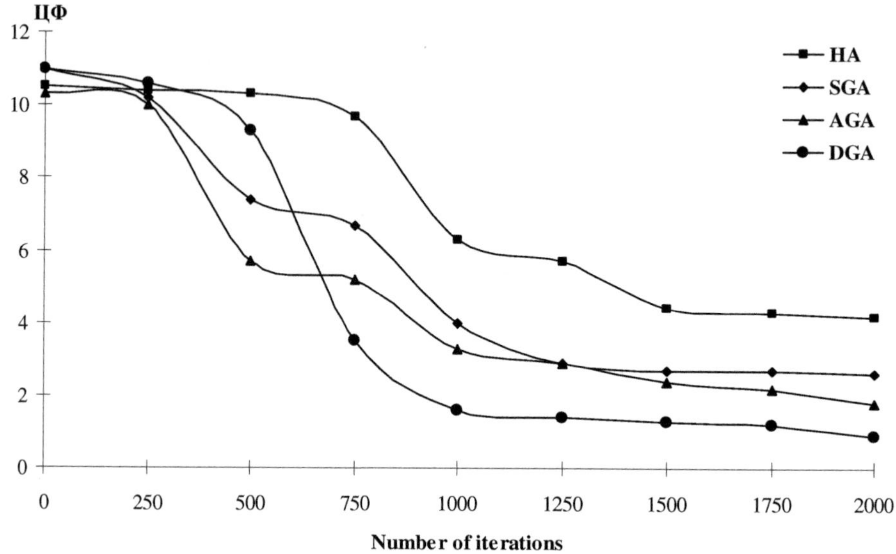

Fig. 5.6. Comparison of performance of tested algorithms

So, the proposed DGA finds a solution almost two times closer to optimum than other algorithms. During initial 500 iterations the algorithm used improves TF with average speed. However, already at around iteration 900 DGA finds an acceptable solution, better than found by other algorithms by the end of the interval assumed.

To determine efficiency of DGA we must compare the algorithms examined in terms of time and memory size used. The respective experimental data are presented in Table 5.9.

Table 5.9. Comparison of investigated algorithms in terms of time and space characteristics

#	Compared algorithms	Solution time, sec.	RAM size, KB.
1	Heuristic algorithm for formula approximation	1169.3361	659.2951
2	Simple genetic algorithm	512.3952	392.6384
3	Adaptive genetic algorithm	281.6843	726.6168
4	Dynamic genetic algorithm	254.5428	409.2886

In this table, the column 'Solution time, sec.' shows average time to problem solution for series of 50 tests. Considering these experimental results we can say that DGA is faster than all examined algorithms, but has somewhat bigger space requirements than SGA.

Considering the results of these investigations we can conclude that DGA provides solutions of better quality than the classical genetic algorithms as it uses the ability to change inoperative algorithm parameters and genetic operators during functioning, thereby preventing useless search for better values of TF in the situations, when the stock of genetic material is exhausted.

5.4 Conclusions

1. Optimal parameter combinations for GAMG have been defined. On the basis of results of experimental investigations of GAMG the parameter values were recommended (population size, selection type, probabilities of crossover, P_k, and mutation, P_m), allowing for obtaining the best performance during search for solution.
2. The space and time complexity of GAMG and DGA were determined experimentally. It was concluded that experimental estimates corresponded appropriately to the theoretical ones.
3. The comparison of GAMG with similar algorithms was performed. GAMG proved to be faster than all the other considered algorithms, but has bigger space requirements than simulated annealing.
4. DGA was also compared with analogous algorithms. This comparison showed high DGA efficiency, as it finds the best solutions of all the algorithms examined. It should be emphasized, as well, that DGA has the highest speed of approaching the optimum solution. Thus, DGA is the fastest of algorithms considered, but it also has bigger memory size requirements than SGA.

Final Conclusions

We can summarize the main results contained in the book as follows:

1. We presented a review and analysis of modern systems meant to provide solutions to engineering design problems.
2. We provided a choice of genetic search and evolution modeling methods for solving problems of automated design of DSD MH. We developed a method of genetic material presentation for algorithms developed by us: the genetic algorithm with multiple genotype and the dynamic genetic algorithm, allowing for reduction of the number of 'illegal' solutions, while keeping high chromosome encoding and decoding speed. This method allows for reducing the time of algorithm work by 10-20% in comparison with known encoding methods.
3. The structures of genetic algorithms for the problems considered have been developed: GAMG for determining optimum MH parameters and characteristics; DGA for determining composition of materials with desired qualities. We defined genetic operators, target functions, self-organization and self-adjustment procedures for these algorithms. We established theoretical estimates of space and time complexity of these algorithms for DSD MH design automation.
4. Software implementations were developed for DSD MH automated design, using the genetic algorithms proposed. In these implementations the object-oriented programming language C++ was used. Software systems display information on TF changes (maximum, average and minimum values) as diagrams; changes of the parent and descendant populations are visualized.
5. Experimental investigations of the algorithms developed have been carried out. Optimal combinations of control parameters of GAMG were found. The DGA parameter changes were analyzed and principles of their self-adjustment established. Empirical estimates of space and time complexity of both algorithms were determined, and comparison of these estimates with theoretical ones was carried out: time complexity is linear for GAMG and polynomial for DGA. These estimates allow for assuming practical value of algorithms developed.
6. Resulting from the investigations of the GAMG search space conclusion was drawn that the algorithm developed was universal in terms of search for optimal values for all the formulated mathematical models of MH parameters and characteristics.
7. Development of new self-organization methods for DGA enabled an improvement in the solution search process, with exclusion of unnecessary iterations when the algorithm finds a local optimum, and automating the setting of parameters of the algorithm. These methods allowed for a significant increase

in the accuracy of parameter setting. It can therefore be concluded that the proposed DGA is a universal algorithm for solving problems of power and factorial complexity.

8. We performed the comparison of algorithms proposed in the book with competitive algorithms from the domain. Using the algorithms proposed allows for reducing time requirements on the development of DSD MH by 10% in comparison with the classical algorithms.

References

1. http://www.tesis.com.ru
2. Bykov, A.V.: CAD/CAM with the distributed functions or possible and impossible automatization miracles. Design automatization 2, 44–47 (1997); In Russian: Быков А. В. CAD/CAM с распределенными функциями или возможные и невозможные чудеса автоматизации. Автоматизация проектирования, 1997, №2, с.44–47
3. Norenkov, I.P.: Genetic algorithms of design and logic problems decision. Information technologies 9 (2000); In Russian: Норенков И. П. Генетические алгоритмы решения проектных и логистических задач//Инфомац. технологии, 2000, №9
4. Artamonov, E.I.: Design of software structures of CAD/CAM systems. Design automatization 2, 56–59 (1997); In Russian: Артамонов Е. И. Проектирование структур программных средств CAD/CAM систем. Автоматизация проектирования, №2, 1997, с.56–59
5. Kureichik, V.V., Kureichik, V.M.: Promising technologies of optimization problems solution. In: Proceedings of the International Scientific Technical Conferences "Intellectual Systems" (IEEE AIS 2003), "Intellectual CAD" (CAD 2003), vol. 1, p. 59. Physmath Publishers, Moscow (2003); In Russian: Курейчик В. В. Курейчик В. М. Перспективные технологии решения оптимизационных задач. В сб.: Труды Международных научно-технических конференций «Интеллектуальные системы (IEEE AIS 2003)» и «Интеллектуальные САПР (CAD-2003)». М.: Изд-во Физико-математической литературы, 2003, Т.1, с.59
6. Kruhzkov, O.A.: On the way to artificial intelligence: critical comprehension of evolutionary foundation. In: The Collection: Proceedings of the International scientific technical Conferences "Intellectual Systems" (IEEE AIS 2003), "Intellectual CAD" (CAD 2003), vol. 1, p. 7. Physmath Publishers, Moscow (2003); In Russian: Кружков О. А. На пути к искусственному интеллекту: критическое осмысление эволюционных оснований. В сб.: Труды Международных научно-технических конференций «Интеллектуальные системы (IEEE AIS-2003)» и «Интеллектуальные САПР (CAD-2003)». М.: Изд-во Физико-математической литературы, 2003, Т.1, с.7
7. Flerov Yu, A., et al.: Automation of design of complex machine-building systems. Design automatization 1, 3–5 (1996); In Russian: Флеров Ю. А. и др. Автоматизация проектирования сложных объектов машиностроения. Автоматизация проектирования, №1, 1996, с.3–5
8. Kureichik, V.M.: Mathematical software of KTP with CAD usage, p. 352. R&C, Moscow (1990); In Russian: Курейчик В. М. Математическое обеспечение КТП с применением САПР. М., Радио и связь, 1990, с. 352

9. Hubka, V.: Theory of technical systems. Mir, Moscow (1987); In Russian: Хубка В. Теория технических систем. М.: Мир, 1987. 289с
10. http://www.synopsys.com
11. Ilyin, V.N.: Application of artificial intelligence in REE CAD. MAU, Moscow (1990); In Russian: Ильин В. Н. Применение искусственного интеллекта в САПР РЭС. М.: МАИ, 1990. 56с
12. http://www.syncad.com
13. http://www.orcad.com
14. Trakhtenghertz, E.A.: Computer support for decision-making. SINTEG, Moscow (1998); In Russian: Трахтенгерц Э. А. Компьютерная поддержка принятия решений. М.: СИНТЕГ, 1998
15. Petrov, A.V. (ed.): CAD development. Vysshaya shkola (1990); In Russian: Разработка САПР. Под ред. А. В. Петрова. М., Высшая школа, 1990
16. Lin, S.C., Goodman, E.P., Punch, W.F.: A genetic algorithm approach to dynamic job shop scheduling problems. In: Proc. of the 7 international conf, pp. 481–488. M. Kaufmann Publisher, San Meteo (1997)
17. Artamonov, E.I.: Complex of GRAFIKA-81 CAD/CAM systems software. Design automatization 1, 42–45 (1997); In Russian: Артамонов Е. И. Комплекс программных средств CAD/CAM- систем ГРАФИКА-81. Автоматизация проектирования, №1, 1997, с.42–45
18. Zamuruyev, A.P.: Pro-ENGINEER - new traditions, modern thinking. Design automatization 2, 54–58 (1998); In Russian: Замуруев А. Р. Pro-ENGINEER – новые традиции, современное мышление. Автоматизация проектирования, №2, 1998, с.54–58
19. Rybakov, A.V., Evdokimov, S.A., Krasnyh, A.A.: Development of systems of automated support for engineering decisions. Design automatization 5, 44–47 (1997); Рыбаков А. В., Евдокимов С. А., Краснов А. А. Создание систем автоматизации поддержки инженерных решений. Автоматизация проектирования, №5, 1997, с.44–47
20. Goldberg, D.E., Kalyanmov, D.A.: Comparative analysis of selection schemes used in genetic algorithms. In: Rawlings, G. (ed.) Foundations of genetic algorithms, Indiana University, p. 190. Morgan Kaufmann, San Mateo (1991)
21. Application of mathematical methods and computers. In: Ostanin, A.N. (ed.): Planning and processing of experimental data, Minsk, Belarus, Vysheyshaya shkola (1989); In Russian: Применение математических методов и ЭВМ. Планирование и обработка результатов эксперимента: Учеб. Пособие. / Под общ. ред. А. Н. Останина. Минск.: Вышэйшая школа, 1989, 218с
22. Darwin, C.: On the Origin of Species by Means of Natural Selection, Works, vol. 3. Academia, M. — St.-P (1939); In Russian: Дарвин Ч. Происхождение видов путем естественного отбора. Соч. т. 3, М. Л.: «Академия», 1939
23. Dubinin, N.P.: The selected works. In: The problems of gene and evolution, vol. 1. Nauka, Moscow (2000); In Russian: Дубинин Н. П. Избранные труды, Т. 1. Проблемы гена и эволюции. М.: Наука, 2000
24. Kureichik, V.V.: Evolutionary methods of solving optimization problems. In: Monograph. TSURE Publishers, Taganrog (1999); In Russian: Курейчик В. В. Эволюционные методы решения оптимизационных задач. Монография. Таганрог: Изд-во ТРТУ, 1999
25. Lamarck, J.B.: Philosophy of Zoology, vol. 1, 2. Academia, M. — St.-P.(1939); In Russian: Ламарк Ж. Б. Философия зоологии. Т. 1, 2, -М. Л.: «Академия», 1939

26. Evolutionary epistemology and logics of social sciences: Karl Popper and his critics. Complied by Lakhuti, D.G., Sadovsky, V.N., Finn, V.K. Editorial URSS (2000); In Russian: Эволюционная эпистемология и логика социальных наук: Карл Поппер и его критики//Составление Д. Г. Лахути, В. Н. Садовского, В. К. Финна. М.: Эдиториал УРСС, 2000.
27. Red'ko, V.G.: Evolutionary cybernetics, M., R. F., Nauka (2001); In Russian: Редько В. Г. Эволюционная кибернетика. М.: Наука, 2001
28. Holland, J.H.: Adaptation in Natural and Artificial Systems: An Introductory Analysis with Application to Biology, Control and Artificial Intelligence. University of Michigan, USA (1975)
29. Goldberg, D.E.: Genetic Algorithms in Search, Optimization and Machine Learning. Addison-Wesley Publishing Company, Inc., Reading (1989)
30. Davis, L. (ed.): Handbook of Genetic Algorithms. Van Nostrand Reinhold, New York (1991)
31. Chambers (ed.): Practical Handbook of Genetic Algorithms, vol. 1. CRC Press, Washington (1995)
32. Chambers (ed.): Practical Handbook of Genetic Algorithms, vol. 2. CRC Press, Washington (1995)
33. Chambers (ed.): Practical Handbook of Genetic Algorithms, vol. 3. CRC Press, Washington (1999)
34. Davis, L.: Genetic Algorithms and Simulated Annealing. Morgan Kaufman Publisher, San Mateo (1987)
35. Koza, J.R.: Genetic Programming. MIT Press, Cambridge (1992)
36. Rastrigin, L.A.: Random Search in Evolutionary Computations. In: Proceedings 1st International conf., Evolutionary Computation and Its Application, EvCA 1996, Moscow, pp. 135–143 (1996)
37. Kureichik, V.V.: Evolutionary, synergetic and homeostatic methods of decisions making. In: Monograph. TSURE Publishers, Taganrog (2001); In Russian: Курейчик В. В. Эволюционные, синергетические и гомеостатические методы принятия решений. Монография. Таганрог: Изд-во ТРТУ, 2001
38. Kureichik, V.M.: Genetic Algorithms. In: Monograph. TSURE Publishers, Taganrog (1998); In Russian: Курейчик В. М. Генетические алгоритмы: Монография. Таганрог: Изд-во ТРТУ, 1998
39. Tarasov, V.B.: From Multiagent Systems to Intellectual organizations: philosophy, psychology, computer science. Editorial URSS, Moscow (2002); In Russian: Тарасов В. Б. От многоагентных систем к интеллектуальным организациям: философия, психология, информатика. — М.: Эдиториал УРСС, 2002
40. Evolutionary calculations & genetic algorithms, compiled by Gudmann E. D., Kovalenko A. P. The review of applied & industrial mathematics. TVP, Moscow (1996); In Russian: Эволюционные вычисления и генетические алгоритмы. Составители Гудман Э. Д., Коваленко А. П. Обозрение прикладной и промышленной математики. М.: Изд-во ТВП, 1996
41. Kureichik, V.M.: Computer Software for Engineering and Technological Design using CAD. R&C, Moscow (1990); In Russian: Курейчик В. М. Математическое обеспечение конструкторского и технологического проектирования с применением САПР. М.: Радио и связь, 1990
42. Genetic Algorithms. In: Elbaum, L. (ed.) Proceedings of the 1st International conf. Associates Publishers, New Jersey (1985)

43. Genetic Algorithms. In: Grefenstette, J. (ed.) Proceedings of the 2nd International conf. Associates Publishers, New Jersey (1987)
44. Genetic Algorithm. In: Schaffer, D. (ed.) Proceedings 3d International conf. Morgan Kaufman Publishers, San Mateo (1989)
45. Genetic Algorithms. In: Belew, R., Booker, L. (eds.) Proceedings of the 4th International conf. Morgan Kaufman Publishers, San Mateo (1991)
46. Genetic Algorithms. In: Forrest, R. (ed.) Proceedings of 5th International conf. Morgan Kaufman Publishers, San Mateo (1993)
47. Genetic Algorithms. In: Forrest, R. (ed.) Proceedings of 6th International conf. Morgan Kaufman Publishers, San Mateo (1995)
48. Genetic Algorithms. In: Back, T. (ed.) Proceedings of the 7th International conf. Morgan Kaufman Publishers, Inc., San Francisco (1997)
49. Genetic Algorithms. In: Goldberg, D. (ed.) Proceedings of the 8th International conf. Morgan Kaufman Publishers, Inc., San Francisco (1999)
50. Potts, C.I., Giddens, T.D., Yadav, S.B.: The Development and Evaluation of an Improved Genetic Algorithm Based on Migration and Artificial Selection. IEEE Trans. on Systems, Man and Cybernetics 24(1), 73–86 (1994)
51. Shahookar, K., Mazmunder, P.: A Genetic Approach to Standard Cell Placement Using Meta-Genetic Parameter Optimization. IEEE Trans. on CAD 9(5), 500–511 (1990)
52. Cohoon, J.P., Paris, W.D.: Genetic Placement. IEEE Trans. on CAD 6(6), 956–964 (1987)
53. Ackley, D.H.: A Connectionist Machine for Genetic Hillclimbing. Kluwer Academic Publishers, Boston (1987)
54. Kuritsky, B.I.: Optimization wherever you look. Mashinostroeniye, St.-P (1989); In Russian: Курицкий Б. Я. Оптимизация вокруг нас. Л.: Машиностроение, 1989
55. Kureichik, V.M., Miagkikh, V.: Some New Features in Genetic Solution of the TSP. In: Proceedings of the second International conference, Plymouth, UK (1996)
56. Nechepurenko, M.I., et al.: Algorithms and programs for solving graph and network problems. Novosibirsk, Nauka, The Sibirian Branch of the R. F. A (1990); In Russian: Нечепуренко М. И. и др. Алгоритмы и программы решения задач на графах и сетях. Новосибирск: Наука. Сиб. отд-ние, 1990
57. Kureichik, V., Tetelbaum, A.: Graph Isomorphism Algorithm for Regular VLSI Structure. In: Proc. 28th Annual conf. in Information Sciences and systems, Princenton, USA, pp. 17–23 (1994)
58. Larichev, O.I.: Theory and Methods of decision-making as well as the Chronicles of events in the Magic Countries. Logos, Moscow (2000); In Russian: Ларичев О. И. Теория и методы принятия решений, а также Хроника событий в Волшебных Странах. М.: Логос, 2000
59. Keeney, R.L., Raiffa, H.: Decision-making with multiple criteria (preferences and substitutions). R&C, Moscow (1981); In Russian: Кини Р. Л., Райфа Х. Принятие решений при многих критериях: предпочтения и замещения. М.: Радио и связь, 1981
60. Bernstein, L.S., Karelin, V.P., Tselykh, A.N.: Models and methods of decision-making in integrated circuits. RSU Publishers, Rostov-on-Don (1999); In Russian: Берштейн Л. С., Карелин В. П., Целых А. Н. Модели и методы принятия решений в интегрированных ИС. Ростов-на-Дону: Изд-во РГУ, 1999

61. Vagin, V.N., Eremeyev, A.P.: Design of intellectual systems of real-time DMP, pp. 27–32. Nauka, Moscow (1999); . In Russian: Вагин В. Н., Еремеев А. П. Конструирование интеллектуальных систем ППР реального времени. Интеллектуальное управление: новые информационные технологии в задачах управления, М.: Наука, 1999, с. 27–32
62. Harary, F.: Graph Theory. Mir, Moscow (1977); In Russian: Харари Ф. Теория графов. М.: Мир, 1977
63. Ore, O.T.: Graph Theory. Nauka, Moscow (1973); In Russian: Оре О. Т. Теория графов. М.: Наука, 1973
64. Cristofides, N.: Graph Theory. In: The algorithmic approach, Mir, Moscow (1978); In Russian: Кристофидес Н. Теория графов. Алгоритмический подход. М.: Мир, 1978
65. Kureichik, V.M.: Discrete Mathematics, part 2. Elements of graph theory. TSURE Publishers, Taganrog (1997); In Russian: Курейчик В. М. Дискретная математика, Часть 2. Элементы теории графов. Таганрог: Изд-во, ТРТУ, 1997
66. Kormen, T., Leizerson, I., Rivest, R.: Algorithms, plotting & analysis. MCMO, Moscow (2000); In Russian: Кормен Т., Лейзерсон И., Ривест Р. Алгоритмы: построения и анализ. М.: МЦМО, 2000
67. Aristotle — Selected works in four volumes, vol. 1. Mysl', Moscow (1976); In Russian: Аристотель. Сочинения в четырех томах. Т. 1. М.: Мысль, 1976
68. Plato — Selected works in four volumes, vol. 1. Mysl', Moscow (1990); In Russian: Платон. Сочинения в четырех томах. Т. 1. М.: Мысль, 1990
69. De Jong, K.: Evolutionary Computation: Recent Development and Open Issues. In: Proceedings of 1st International conf., Evolutionary Computation and Its Application, EvCA 1996, Moscow, pp. 7–18 (1996)
70. Abilov, Y.A., Aliyev, R.A., Nasirov, I.M.: GA with a group selection and directed mutation. The news of Academy of Sciences, Control theory and systems 5, 96–99 (1997); In Russian: Абилов Ю. А., Алиев Р. А., Насиров И. М. ГА с групповым выбором и направленной мутацией//Известия АН. Теория и системы управления, № 5, 1997, с. 96–99
71. Haken, G.: Synergetics. Hierarchy of instabilities in self-organizing systems and devices. Mir, Moscow (1985); In Russian: Хакен Г. Синергетика. Иерархия неустойчивостей в самоорганизующихся системах и устройствах. М.: Мир, 1985
72. Dulnev, G.N.: Introduction into Synergetics. Prospect, St.-Petersburg (1998); . In Russian: Дульнев Г. Н. Введение в синергетику. СПб.: Изд-во «Проспект», 1998
73. Kronover, R.M.: Fractals and chaos in dynamic systems. In: Foundations of theory. Postmarket, Moscow (2000); In Russian: Кроновер Р. М. Фракталы и хаос в динамических системах. Основы теории. М.: Постмаркет, 2000
74. Loskutov, A.Y.: Synergetics and nonlinear dynamics: a new approach to old problems. Synergetics. In: The proceedings of the Siminar, vol. 3, pp. 204–224. MSU Publishers, Moscow (2000); In Russian: Лоскутов А. Ю. Синергетика и нелинейная динамика: новые подходы к старым проблемам. Синергетика// Труды семинара. Том 3. М.: Изд-во МГУ, 2000 с. 204–224
75. Morozov, K.K., et al.: The methods of partitioning the REE circuits into construction-wise stand alone units. Sov. Radio, Moscow (1978); In Russian: Морозов К. К. и др. Методы разбиения схем РЭА на конструктивно законченные части. М.: Советское радио, 1978

76. Karelin, V.P., Rodzin, S.I.: UMP on methods of mathematical programming (search optimization). TSURE Publishers, Taganrog (1990); In Russian: Карелин В. П., Родзин С. И. УМП по методам математического программирования (поисковой оптимизации). Таганрог: Изд-во ТРТУ, 1999
77. Kureichik, V.M., Kureichik, V.V.: A Genetic Algorithms for Graph Partitioning. Journal of Computer and Systems Sciences International 38(4), 580–588 (1999)
78. Frohlich, N., Glockel, V., Fleischmann, J.: A new partitioning method for parallel simulation of VLSI circuits on transistor level. In: Proceedings Design, Automation and Test in Europe Conference, Paris, France, March 27-30, pp. 679–685 (2000)
79. Saab, Y.G., Rao, V.B.: Fast Effective Heuristics for the Graph Bisectioning Problem. IEEE Transaction on CAD 9(1), 91–98 (1990)
80. Wei, Y.C., Cheng, C.K.: A two-level two-way partitioning algorithm, Tech. report CH2924-9, University of California, San Diego. IEEE (1990)
81. Yeh, C.-W., Cheng, C.-K., Lin, T.-T.Y.: A general purpose multiple way partitioning algorithm. In: Proceedings 28th ACM/IEEE Design Automation Conference, paper 25/1, pp. 421–425 (1991)
82. Kernighan, B., Lin, S.: An efficient heuristic procedure for partitioning graphs. Bell Syst. Tech. J. 49, 291–307 (1970)
83. Fiduccia, C., Mattheyses, R.: A linear time heuristics for improving network partitions. In: Proceedings 19th ACM/IEEE Design automation conference, pp. 175–181 (1982)
84. Naveed, S.: Algorithms for VLSI Physical Design Automation. Kluwer Academic Publishers, Dordrecht (1995)
85. Saab, Y.: A new effective and efficient multi-level partitioning algorithm. In: Proceedings Design, Automation and Test in Europe Conference 2000, Paris, France, March 27-30, pp. 112–116 (2000)
86. Bui, T.N., Moon, B.R.: Genetic algorithm and graph partitioning. IEEE Trans. Comput. 45, 841–855 (1996)
87. Chandrasekharam, R., Subhramanian, Chadhurys: Genetic algorithms for node partitioning problem and application in VLSI design. IEEE Proc-E 140(5), 167–178 (1993)
88. Batischev, D.A.: Genetic algorithms of solving optimization problems. VSTU Publishers, Voronehz (1995); In Russian: Батищев Д. А. Генетические алгоритмы решения экстремальных задач. Воронеж: Изд-во ВГТУ, 1995
89. Kureichik, V.M., Kureichik, V.V.: A fractal algorithm of graph partitioning. The news of Academy of Sciences. Theory and control systems. 4, 79–87 (2002); In Russian: Курейчик В. М., Курейчик В. В. Фрактальный алгоритм разбиения графа//Известия АН. Теория и системы управления, № 4, 2002, с. 79–87
90. Kling, R.M., Banerjee, P.: Placement by Simulated Evolution. IEEE Trans. on CAD 8(3), 245–256 (1989)
91. Kling, R.M., Banerjee, P.: Empirical and Theoretical Studies of the Simulated Evolution Method Applied to Standard Cell Placement. IEEE Trans. on CAD 10(10), 1303–1315 (1991)
92. Shahookar, R., Mazumder, P.: VLSI Placement Techniques. ACM Computing Surveys 23(2), 142–220 (1991)
93. Paris, W.: GENIF: A new placement algorithm. Thesis (ms) University of Virginia, USA (1989)
94. Mayer, M.: Parallel GA for the DAG Vertex spelling problem. Thesis, University of Missouri, USA (1993)

95. Kureichik, V.M., Kureichik, V.V.: Genetic Algorithm for Graph Placement. Journal of Computer and Systems Sciences International 39(5), 733–740 (2000)
96. Grefenstette, J., Gopal, G., Rosmaita, B., van Gucht, D.: Genetic algorithms for the traveling salesman problem. In: Grefenstette, J. (ed.) Proc. Intern. Conf. of Genetic Algorithms and their applications, New Jersey, pp. 160–165 (1987)
97. Grefenstette, J.J. (ed.): Genetic Algorithms for Machine Learning. Kluwer Academic Press, USA (1994)
98. Oliver, I., Smith, D., Holland, J.R.: A study of permutation crossing-over operators on the traveling salesman problem. In: Proc. of the Second International Conf. on Genetic Algorithms, pp. 224–230 (1993)
99. Kureichik, V.M., Miagkikh, V., Topchy, A.: Combined Genetic and Local Search Algorithms for the Quadratic Assignment Problem. In: Proceedings, First International Conference on Evolutionary computation and its Applications, Moscow, Russia, pp. 335–341 (1996)
100. Kureichik, V.M.: A Graph Isomorphism Algorithm for Regular VLSI Structures, vol. 3, pp. 4–12. The news of TSURE, Taganrog (1997)
101. Read, R.C., Cornell, D.G.: The graph isomorphism disease. Journal of Graph Theory 1, 339–363 (1977)
102. Kureichik, V.M., Bickart, T.A.: An Isomorphism test for homogeneous graphs. In: Proc. 1979 Conf. on Information Science and Systems, Baltimore, USA, pp. 175–183 (1979)
103. Warnaar, D.B., Chew, M., Olarin, S.: Method for detecting isomorphism in graphs using vertex degree correspondence with partitioning. American Society of Mechanical Engineers, DE 47, 219–224 (1993)
104. Ohlrich, M., Fbeling, C., Cinting, E., Sather, l.: Sub Gemini: Identifying Sub Circuits using Isomorphism Algorithm. In: Proc. 30th DAC, USA, pp. 31–37 (1993)
105. Koza, J.R.: Genetic Programming-2. MIT Press, Cambridge (1998)
106. Kureichik, V.M.: Genetic Algorithms and application. In: Monograph. TSURE Publishers, Taganrog (2002); In Russian: Курейчик В. М. Генетические алгоритмы и их применение: Монография. — Таганрог: Изд-во ТРТУ, 2002
107. Kureichik, V.M., Kureichik, V.V.: A fractal algorithm of graph partitioning. The news of Academy of Sciences, Control theory and systems 4, 65–75 (2002); In Russian: Курейчик В. М. Курейчик В. В. Фрактальный алгоритм разбиения графа //Известия АН. Теория и системы управления, № 4, 2002, с. 65–75
108. Yemelyanov, V.V., Kureichik, V.V., Kureichik, V.M.: Theory and practice of evolutionary simulation. Phyzmath, Moscow (2003); In Russian: Емельянов В. В., Курейчик В. В., Курейчик В. М. Теория и практика эволюционного моделирования. — М.: Физматлит, 2003, 432 с
109. Kureichik, V.M.: Genetic algorithms. Textbook for university students. TSURE Publishers, Taganrog (1998); . In Russian: Курейчик В. М. Генетические алгоритмы. Учебник для вузов. Таганрог, Изд-во ТРТУ, 1998г., 118 с
110. Trudonishin, V.A., Pivovarova, N.V.: Automated design systems: in 9 volumes. In: Norenkov, I.P. (ed.) University textbook. Mathematical models of technical objects, vol. 4, Vysshaya shkola, Moscow (1986); In Russian: Системы автоматизированного проектирования: В 9-ти кн. Кн.4. Математические модели технических объектов: Учеб. пособие для вузов/ В. А. Трудонишин, Н. В. Пивоварова; Под ред. И. П. Норенкова. М.: Высш. шк.,1986. 160с

111. Malyukov, S.P., Obzhelyansky, S.A.: Magnetic head design using CAD. The news of TSURE 3, 226 (2002); In Russian: Малюков С. П., Обжелянский С. А. Проектирование магнитных головок с применением технологии САПР. Известия ТРТУ, 2002, №3, с.226
112. Kolosov, G.E.: On a population size control problem. The news of Academy of Science, Control theory and systems 2, 181–182 (1995); In Russian: Колосов Г. Е. Об одной задаче управления численностью популяции. //Изв. РАН. Теории и системы управления, 1995. №2, с. 181–182
113. Herrera, F., Verdegay, J.L.: Genetic Algorithms and Soft Computing. Physica Verlag, Heidelberg (1996)
114. Gavrilova, T.A., Khoroshevsky, V.F.: Knowledge Bases of Intellectual Systems. Piter, St.-Petersburg (2000); In Russian: Гаврилова Т. А. Хорошевский В. Ф. Базы знаний интеллектуальных систем. СПб.: Питер, 2000
115. Malyukov, S.P., Obzhelyansky, S.A.: Application of genetic algorithms in magnetic head design. Promising Information Technologies and Intellectual Systems (PITIS) 2, 58–64 (2002); In Russian: Малюков С. П., Обжелянский С. А. Применение генетических алгоритмов при разработке магнитных головок. Перспективные Информационные Технологии и Интеллектуальные Системы (ПИТИС), 2002, №2, с.58–64
116. Davis, L.D.: Handbook of genetic algorithms, p. 207. Van Nostrand Reinhold, New York (1991)
117. Chambers (ed.): Practical Handbook of Genetic Algorithms. CRC Press, Washington (1999)
118. Kureichik, V.M.: Genetic algorithms. Review and state-of-the-art,The news of II 3, 14–63 (1998); In Russian: Курейчик В. М. Генетические алгоритмы. Обзор и состояние//Новости ИИ.1998. №3, с.14–63
119. Kureichik, V.M.: Methods of Genetic Search, part 1. Textbook. TSURE Publishers, Taganrog (1998); In Russian: Курейчик В. М. Методы генетического поиска: Ч.1. Учебное пособие. Таганрог: Изд-во ТРТУ, 1998, 118с
120. Davidenko, V.N., et al.: Investigation of genetic optimization methods, Taganrog (1996); In Russian: Давиденко В. Н. и др. Исследование генетических методов оптимизации. //Депонир. в ВНТИ № ГР 02.9.70001.838, Таганрог, 1996
121. Kureichik, V.M., Lebedev, B.K., Nuzhnov, E.B.: Textbook for the course on Genetic optimization methods, Moscow, R.F.; In Russian: Курейчик В. М., Лебедев Б. К. Нужнов Е. В. Учебное пособие по курсу «Генетические методы оптимизации». М., 1996, 132с
122. Blanton, J., Wainwright, R.: Multiple VeLSIle Routing with Time and Capacity Constraints Using Genetic Algorithms. In: Proc. of 5th Int. Conf. on GA. Morgan Kaufmann Publ., San Mateo (1993)
123. Kureichik, V.M., Zinchenko, L.A.: Evolutionary simulation with dynamic modification of parameters. In: Proceedings of the 7th National Artificial Intelligence conference, pp. 516–523. Phyzmath, Moscow (2000); In Russian: Курейчик В. М. Зинченко Л. А. Эволюционное моделирование с динамическим изменением параметров//Труды 7-ой Национальной конференции по искусственному интеллекту. М.: Физматлит, 2000, с.516–523

124. Kureichik, V.M., Malyukov, S.P., Obzhelyansky, S.A.: The structure of internal presentation of data in dynamic genetic algorithm for automating selection of materials with predetermined qualities. The news of TSURE 3(38), 34–38; In Russian: Курейчик В. М., Малюков С. П., Обжелянский С. А. Структура представления внутренних данных динамического генетического алгоритма автоматизации подбора материалов с заданными свойствами. Известия ТРТУ, 2004, №3 (38), c.34–38
125. Kureichik, V.M., Malyukov, S.P., Objelyansky, S.A.: The dynamic genetic algorithm of automated selection of materials with predefined qualities. The news of TSURE 3(38), 38–42 (2004); In Russian: Курейчик В. М., Малюков С. П., Обжелянский С. А. Динамический генетический алгоритм автоматизированного подбора материалов с заданными свойствами. Известия ТРТУ, 2004, №3 (38), c.38–42
126. Kureichik, V.M., Zinchenko, L.A.: Symbolic information technologies in evolutionary modeling. In: Proceedings ECAI 2000, pp. 50–53 (2000)
127. Davis, L.: Handbook of Genetic Algorithms, 412 p. Van Nostrand Reinhold, New York (1991)
128. Cohoon, J., et al.: Distributed Genetic Algorithms for the Flooplain Design Problem. IEEE Trans. on CAD 10(4), 483–492 (1991)
129. Michalewicz, Z.: Genetic Algorithms + Data Structures = Evolution Programs. Springer, Berlin (1999)
130. Koza, J.R.: Genetic Programming: On the Programming of Computers by Means of Natural Selection. MIT Press, Cambridge (1992)
131. Kureichik, V.M., Rodzin, S.I.: Evolutionary algorithms: genetic programming//The news of Academy of Science. The news of Academy of Science 1, 127–137 (2002); In Russian: Курейчик В. М., Родзин С. И. Эволюционные алгоритмы: генетическое программирование. //Известия РАН. ТИСУ, 2002. № 1. c.127–137
132. Koza, J.R., Andre, D.: Parallel Genetic Programming on a Network of Transputers. Technical Report, Stanford Univ. (1995)
133. Kureichik, V.M., Zinchenko, L.A., et al.: Algorithms of evolutionary simulation with dynamic operators. In: Intelligent CAD, pp. 148–153. TSURE Publishers, Taganrog (2001); In Russian: Курейчик В. М., Зинченко Л. А. и др. Алгоритмы эволюционного моделирования с динамическими операторами. Изд-во ТРТУ, Известия ТРТУ, Интеллектуальные САПР, 2001 г., стр.148–153
134. Kureichik, V.M., Malyukov, S.P., Obzhelyansky, S.A.: Self-organization procedures in genetic algorithms. In: Proceedings of international science technical conference IEEE CAD AIS 2004, pp. 20–22 (2004); In Russian: Курейчик В. М., Малюков С. П., Обжелянский С. А. Процедуры самоорганизации в генетических алгоритмах. Труды международной научно-технической конференции
135. Goldberd, D.E.: Genetic Algorithms in Search, Optimization and Machine Learning. Addison-Wesley Publishing Company, Inc., Reading (1989)
136. Norenkov, I.P., Kuzmik, P.K.: Information support of high-tech devices. In: CASL-technologies, pp. 213–217. MSTU Publishers, Moscow (2002); In Russian: Норенков И. П., Кузмик П. К. Информационная поддержка наукоемких изделий. CALS-технологии. М.: Изд-во МГТУ, 2002, стр.213–217

137. Kureichik, V.M., Kureichik, V.V., Malyukov, S.P., Objelyansky, S.A.: The self-adjusting genetic algorithm. In: Proceedings of the Ninth National Artificial Intelligence Conference AIC 2004 (with international participation), pp. 35–38. Tver' (2004); In Russian: Курейчик В. М., Курейчик В. В., Малюков С. П., Обжелянский С. А. Самонастраивающийся генетический алгоритм. Труды Девятой Национальной Конференции По Искусственному Интел-лекту КИИ-2004 (С международным участием), Тверь, 2004, с.35–38
138. Adler, Y.P., Markova, E.V., Granovsky, Y.V.: Experiment planning in the search for optimum conditions. Nauka, Moscow (1971); In Russian: Адлер Ю. П., Маркова Е. В., Грановский Ю. В. Планирование эксперимента при поиске оптимальных условий. М., "Наука", 1971
139. Mushik, E., Muller, P.: Methods of taking technological decisions. Mir, Moscow (1990); In Russian: Мушик Э., Мюллер П. Методы принятия технических решений. М.: Мир, 1990. 205c
140. Rumshinsky, L.Z.: Mathematical processing of the experimental data. Nauka, Moscow (1975); Румшинский Л. З. Математическая обработка результатов эксперимента. – М.: Наука, 1975. – 192 с
141. Lorier, J.-L.: Artificial intelligence systems. Translation from French. Mir, Moscow (1991); In Russian: Лорьер Ж-Л. Системы искусственного интеллекта: Пер. с франц. М.: Мир, 1991. 586c
142. Norenkov, I.P.: Development of automated design systems. University textbook. MSTU Publishers, Moscow (1994); In Russian: Норенков И. П. Разработка систем автоматизированного проектирования. Учебник для вузов. М.: Изд-во МГТУ им. Н. Э. Баумана. 1994. 207с., ил

GPSR Compliance

The European Union's (EU) General Product Safety Regulation (GPSR) is a set of rules that requires consumer products to be safe and our obligations to ensure this.

If you have any concerns about our products, you can contact us on ProductSafety@springernature.com

In case Publisher is established outside the EU, the EU authorized representative is:

Springer Nature Customer Service Center GmbH
Europaplatz 3
69115 Heidelberg, Germany

Batch number: 10620141

Printed by Printforce, the Netherlands